WHY
THE
WORLD
DOESN'T
SEEM
TO
MAKE
SENSE

WHY THE WORLD DOESN'T SEEM TO MAKE SENSE

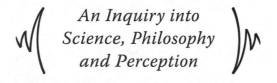

*An Inquiry into
Science, Philosophy
and Perception*

Steve Hagen

SENTIENT PUBLICATIONS

First Sentient Publications edition 2012

First published by Quest Books as *How the World Can Be the Way It Is,* 1995

Cover and page 212: M.C. Escher's "Up and Down" © 2011 The M.C. Escher Company-Holland. All rights reserved. www.mcescher.com

A paperback original

Cover design by Kim Johansen
Book design by Timm Bryson

Library of Congress Cataloging-in-Publication Data
Hagen, Steve, 1945-
  Why the world doesn't seem to make sense : an inquiry into science, philosophy, and perception / Steve Hagen. -- 1st Sentient Publications ed.
    p. cm.
  Previous edition published: How the world can be the way it is. Wheaton, Ill., U.S.A. : Quest Books, 1995.
  Includes bibliographical references and index.
  ISBN 978-1-59181-180-0
  1. Theosophy. 2. Science--Philosophy. 3. Perception. I. Hagen, Steve, 1945- How the world can be the way it is. II. Title.
  BP573.S35H34 2012
  299'.934--dc23
                    2012026767

Printed in the United States of America

10 9 8 7 6 5 4 3 2 1

SENTIENT PUBLICATIONS
A Limited Liability Company
1113 Spruce Street
Boulder, CO 80302
www.sentientpublications.com

IN MEMORY OF

Dale C. Hagen

BELOVED BROTHER, MENTOR, AND FRIEND

⅏(CONTENTS)⅏

PART I
NOBODY KNOWS
WHAT'S GOING ON

PART III
WHAT MATTERS

CHAPTER NINE - BECOMING 240

CHAPTER TEN - TOTALITY 258

EPILOGUE 267

PREFACE TO THE SECOND EDITION

I first published this book in 1995, under the title *How the World Can Be the Way It Is*.

After seventeen years, however, I felt it needed considerable revising. Times have changed and science has made many new discoveries.

Despite much progress elsewhere in those seventeen years, though, science has made no progress in the study of consciousness, which remains as much a mystery to brain researchers as ever.

The original edition of this book had much to say about consciousness that science continues to overlook—including the fact that consciousness itself is not actually a mystery. These insights also form the heart of this second edition. This revision, however, has given me the opportunity to clarify a few critical areas, to make some improvements in terminology, and to update certain sections and passages. In addition, with the help of Scott Edelstein, I have reworked the entire text for clarity.

When people referred to the title of the original edition, they often got it wrong. So I have changed the title to make it both easier to recall and more accurate.

When the first edition of this book appeared in 1995, I was already at work on a sequel. That volume, as of 2012, is at last nearing completion.

⤏{ ACKNOWLEDGMENTS }⤎

Thanks to Scott Edelstein, my longtime friend, editor, and literary agent, for his invaluable help in rewriting this fairly extensive revision of my first book, *How the World Can Be the Way It Is.* This project would not have gotten off the ground without Scott. Thanks also to Jose Palmieri for producing the Newton fractals and Mandelbrot sets, and for reproducing all of the line drawings that appear in this volume. Thanks, too, to Mary Ann Forester for the original drawings of Bohm's fish and Schrödinger's Cat. And finally, thanks once again to everyone else who helped in various ways to put together and publish the original manuscript.

All phenomena, existing and apparent,
Are ever transient, changing and unstable;
But more especially the worldly life
Hath no Reality, no permanent gain.
And so, instead of doing work that's profitless,
The Truth Divine I'll seek.

To the Dragkar-Taso Cave I'll go,
to practice meditation.

—MILAREPA

⊰(INTRODUCTION)⊱

A thing is what it is. What could be clearer? Could there be a proposition more convincing? How could anyone doubt it? And could there be any greater absurdity than that a thing is what it is not?

Yet things being what they are—which is just what common sense would have us believe—results in a world which makes so very little sense. For if things are what they are, then there could be no becoming, and the world, contrary to experience, must be immutable, sterile and devoid of consciousness.

PARADOX AND CONFUSION

I watched the sun rise out of Minnesota, and I followed it. It came down through orange sky and set among the silhouettes of odd angled flat roofs in a jumble of buildings.

The next day, when it rose again, I took the fast train west from Tokyo, past Mount Fuji, past newly-planted terraced fields of rice—past things I didn't know or understand yet, but which seemed familiar in their strangeness. Everything was hot and steamy wet.

A friend's question echoed in my mind: "Why would a Lutheran boy from the Midwest go to a Zen monastery?" I added to myself, "What am I doing here?"

I ferried to the island; I went back in time.

By sundown I was climbing the mountain path up a narrowing gorge. Five times the path crossed the river over frail wooden bridges. I met no one on the path.

I paused at the last bridge, for here the gorge had widened and the river collected in a pool of cold, clear, refreshing water. Not far off, through trees, the floor of this little gorge ended abruptly at a rock wall. I could see where steps had been cut into the side of the mountain. First they traversed the rock face; then, above the wall, they turned from the river and rose in serpentine flights through trees and out of sight.

Once again, in silence—in scented, moist, hot air that stirred vague memories—I lifted my pack, took a deep breath, and continued the climb. In the gathering dark, I occasionally knocked my pack against the rock.

The final ascent was straight up a long run of stairs to an enormous gatehouse. Beneath the upturned roof, two giant figures stood in silhouette against the last glimmer of twilight. Even at this distance, I saw that they were very large, and that they stood with menacing aspect.

As I approached they appeared wind-blown, for their robes seemed to fly about them—yet there wasn't any wind and nothing was moving. I could see their eyes, fierce and fixed upon the stairs. They stood frozen in hideous expressions, looking down upon me as I came. With snarled lips, one's mouth was drawn back over bared teeth in a silent scream; the other, with mouth shut and down-turned, frowned with serious intent. They were like two gigantic ghost-demons, now glowing in the light of the rising full moon.

Yet I came on.

Quickly, silently, the small figure of a monk appeared between them. He had come to take my bag and to usher me in. "Who are they?" I asked. "Paradox and Confusion," came the reply, "the guardians of Truth."

THE PROBLEM

Why was I here? I asked myself this question repeatedly for days. Yet I knew full well why I had come. Things did not add up and I wanted to know why. There was something enigmatic about the human world. For me it all seemed so pointless. After years of study in various fields, after modestly pursuing several careers—what did it all mean when there was no "ground floor" to what I was doing? "So what?" was the answer I always came to after long investigation of any question. Yet I was unwilling to sell out to convenience, to ignore paradox and confusion in order to merely feel comfortable. What comfort could I hope to find when my mind would not be at ease?

It has been said that no tragedy can be written about our time, for there are no more great kings or rulers who can make the tragic mistake which thereby condemns a people. Yet it seems to me that there is much in our day that is genuinely tragic. But the tragedy of our time doesn't lie in any singular event, for it is more like a colorless decline—not unlike that of, say, ecological stability in the face of global warming.

The tragedy of our time is that the light of wisdom has gone out of us, and that even our great institutions of religion and science—not to mention politics, economics, psychology, and philosophy—can no longer save us.

Something is tragically wrong with the human world. I felt sure I was not alone in thinking this. Many people sense that disaster is pending, that quality has left, that life itself is becoming devoid of meaning. There is hardly an educated person anywhere who does not sense in their bones the possibility that we're rushing headlong toward some great calamity—if not a calamity in the physical environment, then at least a calamity for the human spirit. There are no longer any myths to bolster us, and we seem, despite all our efforts, only to accelerate the pace toward our demise. We seem, at times, to not know what we are doing, and yet we are doing a great deal and we're doing it at breakneck speed.

Why this apparent madness to human life, I wanted to know? I wanted to get to the bottom of things. I wanted to get to the Truth of human life—even though, as I was soon to find, it meant going beyond Paradox and Confusion, to letting go of all I knew or believed.

————————————

My teacher pointed to the stepping-stone in the garden. I looked and saw what I assumed was shallow flagstone lying upon the ground. "Stone deep," he said. "You step. Feels good."

It did feel good. The good feeling, of course, was firmness and solidity under foot. This is just what all of us desire. Above all, we want something solid to stand on. Something true. Something real.

Advertising people understand this point. That's why they used to tell us "Coke is the real thing." It must have been a very successful ad because it was with us for quite a while. It really spoke to the bottom of our minds, because what we all want, above all, is the Real Thing.

But we're usually very confused about what that might be. What we think we want today we soon forget about—and so, on the surface, we act as if we don't want the Real Thing at all but, rather, fads, glitter, and the superficial. We place the shallow slab in the garden because we don't fully appreciate our strong desire for deep-set stone—or perhaps because we assume there can be no deep-set stone.

But our confusion is not that we do not *see*. It is more subtle than that. It is that we don't realize what we *see*, and this adds a paradoxical spin to our experience. On the whole, we're not given to noticing that, no matter who we are, scientist or aborigine, we're very prone to story-telling—and to listening only to our own story.

In fact, rarely do we realize that our story is just one among many. Indeed, for most of us, our private story readout is so smooth that any awareness of "this is a story" never registers at all. Jeremy Campbell, in his book *The Improbable Machine*, gives an eloquent account of how we often form our stories out of "flimsy or contradictory data," yet each of us locks on to our story and spontaneously re-makes its explanation, even "in the face of devastating logical argument."[1] It seems that, once we have a story, we're not given to re-examining the evidence for it.

Why do we not simply, naturally and automatically move toward that which is deep-set and Real?

We do not because, when our attention is drawn toward the Real Thing, we meet with thoroughgoing Paradox and Confusion, and we become frightened. We fear to account for what we perceive directly, prior to (and outside of) our story.

Actually, there is no "prior to" for most of us in most circumstances. We're too quick with our story. If we're not quick enough, however, especially when something totally unexpected gets our attention, then we may sense a bit of confusion. Unfortunately, however, we're very quick to manufacture a new, revised story right on the spot. As Campbell puts it, "the mind excels in its effortless ability to treat the world as if nothing it contains is entirely strange." We compulsively "interpret what is new in terms of what we already know."[2] We prefer to do this rather than fully attend to what we directly perceive.

This fear we have of slipping into confusion was profoundly expressed by a student who came to me for meditation instruction a few years ago. He feared he would have a severe collision with Reality if he meditated. He was afraid that he would "disappear," as if at the end of the meditation I would find his unclaimed body lying on the floor.

This is just how we fear Truth. It is nothing less than the fear of the loss of identity.

So, while we long for the Real Thing, we also fear to draw near to it, for the approach to Truth confuses us. And in our fear and discomfort, we begin to mistake one thing for another.

WHAT'S IT ALL ABOUT?

We often say things like, "this means that," yet rarely, or never, do we pause to question if it is ever possible that *this* could truly mean "that." Indeed, it is difficult for us to recognize that equating one thing with another, or making one thing mean another, could pose any serious problem in our thinking at all. But it does, as we shall see. In fact, it is morally devastating.

We've grown so accustomed to our inability to recognize things for what they are (that is, as they would actually appear prior to our conceptualizing them) that we mistake what cannot possibly be Real for Reality. But it's an easy mistake.

A few years ago, at a farmers' market, I saw an enormous gourd sitting on the ground. I didn't know what it was, but I was intrigued by its great size. When I asked the woman behind the counter what it was, she said, "Oh, that's a banana squash." I felt a moderate sense of relief, and my question vanished—that is, my slight discomfort, my "?" state of mind vanished.

Shortly thereafter, however, while wandering through the market, I again came upon the gourd. I suddenly realized that I still didn't know what it was any more than I did before the woman supplied me with a name. All I knew now was that this woman and I would agree to call this thing, whatever it was, a "banana squash." I managed to stop myself just prior to acquiring the belief that I understood something that in fact I did not.

We human beings are commonly confused about appearance and Reality, about identity, about what we really desire, and about what we can and truly do *know*. In short, we are confused about Reality. We've formed many a theory and belief, but as we look about the human world, it is quite clear to us that nobody actually knows what's going on. Yet claims to Truth are being made at every hand, including the claim that there is no Truth.

All of this suggests that all our speculative thoughts are nothing more than conceptual constructs. Since we've fashioned these constructs out of our ignorance, they can reflect nothing conclusive about deep Reality. If there is meaning behind the word "truth," it is quite clear that most of us have taken the wrong approach to finding it.

In this book we'll consider just how deep and how subtle our story lines can go—and in every case they go all the way to Paradox and Confusion.

If you rarely encounter paradox in your daily life, then you simply have not pursued your beliefs all the way to their bottom. We rarely follow our beliefs this far, but paradox awaits us all if we posit any Absolute (which, as we shall *see*, is what we repeatedly and unwittingly do with belief).

We do not appreciate what we actually **do** *see* (i.e., perceive what bare attention reveals to us). Rather, we go with what we think, with what we believe; we go with concepts. At best, however, concepts and beliefs can yield only partial, temporary glimpses of a changing, relative world.

Joseph Campbell suggested that we moderns could use a new myth, but it must be a myth that we can believe. Unfortunately for us, however, we seem to have all but exhausted any possibility for holding onto any objective belief—that is, belief in a "that, out there." The magic and mystery of nature are gone for us. We do not see great significance— let alone deities—in every object, every place, and every dream. In other words, God is not a convincing argument for many of us. Meanwhile, our endeavors to study the external world have made it into virtually lifeless matter for us. We even wonder if we, too, are lifeless—that is, soulless. When everything, including the self, becomes a cold, lifeless object to the senses, the mind can be seen as lifeless as well. "Isn't the mind really a machine?" we ask. "A computer, perhaps? A bit of biological programming locked into protoplasmic hardware?"

These questions lead us inevitably to other, larger ones: What are we? What are we doing here? What is going on? And how can the world be the way it is?

The world is indeed a strange place—inconceivably strange. To briefly illustrate just how strange, we can imagine listening in on the following conversation between a physicist and a philosopher:

> *Physicist:* ...and so we conclude an electron is a particle.
> *Philosopher:* But you also claim an electron is a wave.
> *Physicist:* Yes, it's also a wave.
> *Philosopher:* But surely, not if it's a particle.
> *Physicist:* We say it's both wave **and** particle.
> *Philosopher:* But that's a contradiction, obviously.

Such discoveries in physics—of contradictions that seem to be woven into the very fabric of physical reality—have shattered all our common-sense notions of how the world is made, and of how it works.

In fact, some of these discoveries have been so disconcerting that in October 1987, a group of prominent physicists and philosophers convened at the University of Notre Dame to discuss "the philosophical lessons from quantum theory." Their ultimate goal was to "construct a framework that is empirically adequate, that explains the outcomes of our observations, and that finally produces in us a sense of understanding how the world can be the way it is."[3]

Decades later, such understanding continues to elude us—for even as empirical research brings more and more of our commonsense beliefs under scrutiny, both modern science and modern philosophy have failed to come to grips with our most basic assumptions about the nature of Reality. Nor has either been able to suggest any practical or ethical guidelines that mesh with the quantum reality that science has discovered. In short, it would seem we have no idea of what constitutes Reality. And without any idea of what constitutes Reality, we can have no clear idea of what constitutes Knowledge. The conversation between the physicist and the philosopher might, therefore, continue like this:

Physicist: Are you then saying it's neither wave nor particle?
Philosopher: No, I'm asking what you mean by "it."

If we would carefully examine our experience of our everyday world in light of the philosopher's question, we will soon realize that we are also faced with a knowledge crisis—for if, when we carefully examine the fabric of physical reality we have no way of finally conceiving that reality, then our knowledge of what is truly Real must come into question. And if we don't know what is Real, then in what sense can we say we have any True Knowledge at all? How can we distinguish knowledge from mere belief? The prevailing discourse on knowledge leaves us at an impasse. The consensus is that knowledge **must** be other than mere belief, yet Western philosophy has not been able (or has any idea how) to define knowledge in a way that genuinely distinguishes it from mere belief.

Ultimately, this situation leaves us with a crisis in ethics as well, for without Knowledge of Truth and Reality, we possess no apparent means to resolve the great moral questions that have plagued humankind. Given the above conditions, we have no basis for morality whatsoever. And this is where humankind's woeful state of ignorance begins to pierce the heart and tear at the fabric of society—for our problems with ontology and epistemology are not mere abstractions confined to some seemingly remote quantum world. The problems we face are as close as our social interactions, our everyday lives. Indeed, they are as close as the most intimate, inner workings of our own minds.

In this book, we shall explore these questions—and, in doing so, we'll discover that all of these difficulties are of a single nature. They are all forms of what the ancient Greeks called the paradoxes of "the one and the many"—specifically, the "paradox of plurality." (I will refer to it here simply as the "two-not-two" paradox.) This paradox permeates and underlies virtually all of our mental experience.

The problems in our thinking that stem from this paradox are intensely meaningful for us, for virtually everything of human value is supported by our notions of what is Good, Real and True—the very things this paradox prevents us from getting in touch with.

As we will see, however, contradictions of the two-not-two kind are not in the world itself, but in how we package the world in our minds. When we package the world in conceptual thought, contradiction will always appear. Indeed, as we shall see, we cannot hold the world in **any** conceptual framework without also holding a contradiction.

The two-not-two paradox is extremely subtle, yet it pervades virtually all of human experience resulting from conceptual thought. Thus it has generated many of the great, seemingly intractable problems of Western philosophy. This paradox not only accounts for the appearance of the wave/particle duality in physics, it also accounts for such long-standing problems in philosophy such as free will vs. determinism, the mind/body problem, the organism/environment problem, the problem of appearance vs. reality, the problem of "other minds," etc. It even accounts for the "brother's keeper" dilemma in ethics. Indeed, our confused response to the two-not-two paradox fosters many, if not most, of the social and political ills of humankind.

It is through reliance on perception, rather than conception, that we have an opportunity to resolve this essential paradox, and through which we can find an effective moral, philosophical and psychological framework for living our lives. But, as we shall see, we're commonly confused about conception, and habitually mistake it for perception.

The primary purpose of this book is to help us learn to perceive the world directly—as it is, and not merely as how we conceive it to be. Ultimately, it's only through learning to recognize—and through learning

Note to the Reader

There are two truths—one relative, conditional, changing, bogus; the other Absolute and unmoving. The first refers to our everyday world of relationships—the relative world of things and ideas. The other ("Truth" with a capital T) refers to Absolute—not "THE ABSOLUTE," or "an Absolute," but simply Absolute.

The convention of capitalizing Absolute while leaving the relative in lower-case will be extended in this book to terms referencing Absolute as well, such as Reality, Truth, Knowledge, etc. Thus lower-case terms such as "knowledge" or "seeing" will continue to be used in the common way—that is, in reference to ideas or objects.

Our problems stem from the fact that these conceptual terms cannot refer to Truth or Reality. The term "knowledge," for example, as we shall see, is indistinguishable from what we otherwise call belief when used in the conventional way. Actual Knowledge is quite another matter, however, and must be indicated accordingly. More on this later.

References to Awareness of Truth and Reality, such as knowing or seeing, will be indicated by the use of italics, whereas to rely on—perception that each of us can answer profound moral questions, resolve philosophical and ethical dilemmas, and live lives of harmony and joy.

In Part I we'll see that contradiction necessarily lies at the heart of all our commonsense views of Reality, no matter how we construct them. In Parts II and III we'll see how this revelation may help us reverse our otherwise painful and relentless march from innocence to ignorance. We may find a new innocence, however: an innocence that comes from having exhausted knowledge—the innocence born of wisdom.

But first we must examine how it is that we move, not just from childhood to adulthood but, culturally, from innocence to ignorance. We will consider what it means to be, to *know*, and to assert something. We will

Note to the Reader (continued)

conventional uses of these terms—without italics—will be reserved to refer to the knowing or seeing of thoughts, feelings, ideas or objects in the conventional sense.

Given this specific use of capitalization and italics in this text (italics will be used to introduce the occasional foreign term as well), emphasis of particular terms or phrases will be expressed in **bold text**—which is really what the eye suggests, anyway. Glance at a page of text and it will be the terms in **bold** that jump out at you.

What I have to say in this book is anything but linear—everything in it happens, as it were, at once. Each section relies not only on what comes before, but on what comes after as well, much like a circle of people sitting on one another's laps. Since language is linear, however, the only way to apprehend what I have to say is one sentence at a time.

There are, though, other ways of "getting" what I'm talking about than by reading a book.

explore what we can actually experience directly—and, as we will *see*, this is not at all what our common sense would have us believe we experience. Our method of inquiry will be strictly empirical, yet without any reference to any belief structure (such as we find in, say, science, philosophy, politics, or religion). This will lead us to examine our beliefs, which in turn will force us to look at belief itself, and at our basis for knowledge.

We'll consider how we use belief and how we rely upon it, only to suffer as a result. We'll then consider how belief differs from actual Knowledge and how we are frequently confused about this difference. Finally, we'll explore how Knowledge (i.e., direct perception) in fact reveals a world of sheer wonder and magic that, though transcendent of meaning, is robustly moral.

Our journey will then have taken us full circle. In Part III we will confront and resolve the reality crisis, and arrive at a workable resolution to our chronic problem of everyday life—i.e., what's it all about? As we shall see, it's not in what we are able to think. Rather, it's in what we are able to *see*.

part I

NOBODY KNOWS WHAT'S GOING ON

Hui-neng, the sixth Ancestor, asked: "Whence do you come?"
  Huai-jang of Nan-yueh said: "I come from Tung-shan."
  "What is it that thus comes?"
  Nan-yueh did not know what to answer. For eight long years he pondered the question; then one day it dawned upon him, and he exclaimed, "Even to say it is something does not hit the mark."

one

 BELIEF

*I have discovered that it is necessary, absolutely
necessary, to believe in nothing. That is, we have
to believe in something which has no form and
no color—something which exists before all
forms and colors appear.*

—SHUNRYU SUZUKI

*For it is sufficient, I think, to live by experience,
and without subscribing to beliefs...*

—SEXTUS EMPIRICUS

THE TROUBLE WITH BELIEVING

When I was a child I lived on the side of a hill. It was a broad, grassy escarpment, furrowed by wooded gullies and pierced by outcrops of gabbro, and it rose high behind the houses of my neighborhood. I was told by adults that unicorns romped in those hills.

I don't think I ever believed this, however. My friends and I often hiked there, and we never saw any unicorns. Besides, there was something in the eyes of the adults—a bit of glee, perhaps—that made them seem less than convinced of their own story.

The question of unicorns, of course, was never a serious one. But I was told other things—things that, even to the adults, were clearly not meant to be far-fetched. These stories were not so easily dispelled, for many people believed them. And I used to wonder, what was required for me to believe?

I was raised in a strict Christian home where religion was a daily matter of serious concern. I was brought up to believe that Christ was my personal Savior. This belief was of extreme importance to me as a child, because I was told that in order to be saved from eternal damnation, all I had to do was to believe in Jesus.

Well, I certainly did not want to be damned for eternity, so I was very motivated to believe what I had been told. But there was something enigmatic about the proposition. I wasn't sure just what it was I was supposed to do—that is, **will** myself to do. What was my responsibility? If belief was, as it seemed to be, a moral question, what was I to be held accountable for? As it was presented to me it seemed rather easy. "Just believe," I was told, "and you'll be saved." **Just** believe—but what could this possibly mean? Surely not just to **say** that I believed.

The incident that brought this matter to a head occurred when I discovered that my church frowned on the idea of evolution. I had, by the age of twelve, become convinced through my readings that the theory of evolution explained clearly how life occurred and developed on this planet. Suddenly, I discovered that my belief regarding evolution was in direct conflict with my religious instruction. I was in a quandary, for I did not wish to be damned, yet I could not **choose** to believe as my church would have me believe. I didn't know what to do.

I knew, if I was to be honest with myself (and I was taught, and believed in, the importance of being honest), that deep down I truly did **not** believe. To believe in creationism, I would have had to dismiss other things from my mind that I already knew (believed, really) and understood as valid. I was not at liberty to simply start believing any notion that others happened to declare was true, correct, proper or necessary. In other words, I was powerless to make myself believe what I—or others—would want me to believe.

What does it mean to believe, to hold an opinion? It certainly doesn't mean merely that we say so. It means something very deep. It must mean that deep down we think that such and such is True. It refers to a state of mind that we are powerless to choose.

HOW OUR BELIEFS CHANGE

Since our beliefs do change from time to time, how do we acquire new beliefs if we are not able to simply choose them?

Let's consider this example, which might reveal more about the dynamics of believing. When Albert Einstein published his general theory of relativity in 1915, he didn't believe that the universe was expanding, even though certain parts of his theory suggested that it was. He thought that it was just some strange quirk in the math that implied an expanding universe, so by inserting a special term he was able to get rid of the troublesome part that predicted the expansion, which he regarded as an absurdity.[1]

Then, in the 1920s, astronomers discovered that the distant galaxies were receding rapidly from the Earth—and the more distant the galaxy, the greater its speed of recession. This was an observed fact—physical evidence that implied an expanding universe. Einstein later referred to his alteration of his theory as the biggest blunder of his career.

The point that concerns us here is that, since the 1920s, Einstein believed that the universe might very well be expanding. He didn't believe that in 1915. In 1915 he thought the idea so absurd that he rejected it in his theory. His mind later changed, but how did that happen? It didn't change because he simply chose to change his mind. He didn't just sit down one day and decide to believe that the universe was expanding.

Rather, his mind was changed because it was overwhelmed by a new awareness. In the moment in which he became aware of something new, his mind was different. It had changed—automatically, we could say, for at no point did volition ever directly enter into this change of mind. In fact, it never does.

CONFUSING WHAT WE BELIEVE WITH TRUTH

We should not be too alarmed that our will cannot directly determine what we believe. There is nothing inherent in our beliefs that make them True, and this applies whether or not counter beliefs are (or could be) offered. For example, if you believe that everyone likes spaghetti, there is nothing about that belief which validates that proposition; but, more than that, such lack of validation would remain even if no one could find any contrary evidence.

When it comes to spaghetti, of course, what we believe does not pose a serious problem (unless we're inordinately wild about pasta); but in matters of seemingly greater social consequence, such as questions regarding Truth or moral conduct, relying on what we merely believe could prove to be a needless and disastrous mistake.

People have always believed devoutly, even unto war and death, in a great variety of things that have nothing Real to validate them. Indeed, the very fact that we don't all believe the same things is enough to draw all beliefs into question.

One way of attempting to get around this problem is to grant the legitimacy of all beliefs. This would be very democratic, and perhaps conducive to peace among people of varying views. But, while this arrangement could possibly lead to a more peaceful world, it would nevertheless do nothing to show us Truth and Reality.

Let's consider, briefly, just how problematic this issue can become. Suppose two people hold contrary views regarding some phenomenon—such as, say, the object pictured in Figure 1–1. Though each sees it independently, if they enter into a discussion about their independent

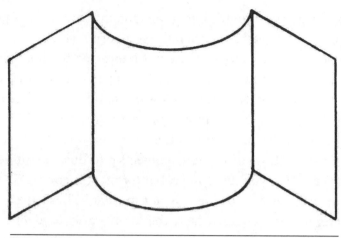

Figure 1–1

objects, it's quite likely that they will believe they are talking about one and the same thing—namely, a single entity "out there" that exists in Reality, that possesses an identity unto itself, and that therefore can be objectified as one and the same for all observers. In this case, however, one person sees a concave surface illustrated, while the other sees it as convex.

Let's say that one of these people insists that only one view is possible—say, that Figure 1–1 **is** convex, and can **only** be convex. Their taking this stance indicates that they do not have a strong feeling for how appearances can differ from Reality. "The way I see it is the way it is," typifies this view.

This view causes problems when it is applied to this simple drawing; when such a view is applied to moral questions—e.g., when does human life begin, or when is war justified—the immediate and inevitable result is conflict, often anger, and sometimes violence.

It is not difficult to see that Figure 1–1 can in fact be viewed in different ways. For many of us, it is equally easy to see how certain moral questions can also be approached and seen in more than one manner. The person who can see how a moral issue can be viewed in different ways is likely to appear quite liberal, tolerant, magnanimous, open, and perhaps even more intelligent than the stubborn dogmatist.

Many philosophers, however, are not impressed by those who take this more generous view. Their reason is that while the first view—that of the dogmatist—clearly does not allow one to fully appreciate the subtle and potentially troubling ambiguities of life, the second view simply makes the morally weak assertion that "for you, it is like that, but for me, it is like this." If we insert actual objects into this view, it begins to appear absurd. "For you it is convex but for me it is concave." "For you abortion is murder but for me it is not murder." These are not compelling moral arguments. In fact, they only carry us deeper into a morass of endless questioning where, as in the case of abortion, we soon begin to discuss not just patently absurd, but inevitably dead questions: Is a fetus human? When does it become human? We may as well debate seriously the question regarding the primacy of the chicken or the egg, or how many angels can dance on the head of a pin. In very short order we are lost and drifting without a mooring.

Philosophers have pointed out that such arguments carry us into a morass because (1) either we would have to assert that the object under discussion **is** both ways in Reality (a paradox), or (2), we end up saying that everything is to each person such as it is to that person (a tautology). We find in either case that we haven't really described anything meaningful.

We can't seem to get to the bottom of things. Though it seems quite evident that there are such distinctions to be made, when we try to pin down just what it is that we mean by them, we find we can't do it. We never seem to touch upon any reality that is distinguishable from appearance—and appearances, we have learned, often deceive. We seem unable to clarify this dense problem of Reality. We are, in short, very prone to falling back into the realm of the dogmatists.

We may argue that at least when we grant the legitimacy of the views of others, we will not be contributing to the hatred and fighting that have gone on since time immemorial among peoples of differing beliefs. But at what cost must we hold such a view? Where does this seemingly more magnanimous view get us? When it comes to the profound questions of human life and moral conduct, we find that we only sink deeper into the mire of human misery. Where at least the dogmatist has something to believe in that gives them a sense of purpose (however deluded it may be), the relativist is without even this and is instead left vulnerable to a feeling of utter meaninglessness.

This sense of meaninglessness is a prevalent problem among intelligent, educated people these days. When carried to the extreme, it reduces the individual to either (1) becoming a zealot out of desperation and thereby sacrificing their intellectual integrity; or (2) living with the sense that ultimately there is no hope or meaning to human life.

In this book we will explore a third possibility that lies outside these two wretched alternatives of utter ignorance or utter despair. By this I do not mean to propose the weak alternative of becoming, in some manner, a fake—i.e., someone who doesn't believe but who pretends to, or who does believe but pretends not to. All such positions are bound by belief, and that is our chronic problem. Rather, I propose that we learn to face our background ignorance squarely and *just see* its nature. We must simply learn to *just see* what is actually presented in experience rather than rely upon **any** belief whatsoever. In so doing we'll be able to *see* that it is precisely in holding our beliefs that we make a profound conceptual mistake. We must learn to rely solely on what we *see* rather than upon what we think.

TRUTH CANNOT BE AN OBJECT

Why can't our beliefs carry us beyond the relative world and lead us to Truth? One reason is that, being relative things ourselves, we cannot leave the relative world. But the question I'm posing here more specifically asks why the act of believing will not allow us to somehow transcend the mundane world of things and ideas.

Let's take a closer look at the dynamics of believing. Consider how the way in which we think, speak and act frequently follows from our innate desire to get other people to believe as we believe. We proselytize others because it makes us feel better. And the reason it makes us feel better is because we're unsure of what we believe ourselves. The pain of this uncertainty becomes less conspicuous to us if we can lose ourselves among others of similar beliefs.

We're unsure of our beliefs because whatever we happen to believe about anything can never provide us with Certitude. As we shall see shortly, all beliefs, since they are conceptual, are necessarily relative and subject to change. They do not, and cannot, provide us with the solid ground we desire. Indeed, our beliefs are actually a source of anxiety.

Having a ground, however, isn't necessary for getting to Truth—in fact it's a hindrance. While we do need our stepping-stones, our beliefs, to get along in the relative world, we **don't** need them in our search for Truth. Indeed, they have no business there. They will only serve to cloud our vision and supply us with false views.

What I'm discussing here is analogous to the enormous problem encountered by mathematicians in the early part of the last century. David Hilbert, perhaps the foremost mathematician of the time, set forth at the International Congress of 1900 nothing less than a proposal to find a way to put mathematics back on a solid footing—a move that led to the development of the formalist school. Even today most people believe that mathematics, being the most exact of all sciences, rests upon solid ground. But this is not the case.[2] Certain anomalies had crept into the growing body of mathematics over the centuries, and by the end of the nineteenth century the philosophical underpinnings of mathematics were beginning to crumble. It was remarked that mathematics was beginning to appear like a fifty-story building erected from the twentieth floor on up. By 1900 mathematicians had become very concerned that there was no solid foundation beneath the edifice that they had built and were continuing to build upon. It was precisely this problem that Hilbert sought to correct.

In 1931, however, all hope of ever putting a solid foundation into place was dashed by Kurt Gödel when he published his two theorems, which proved that any system of mathematics powerful enough to do arithmetic could not be both consistent **and** complete. Gödel managed to set up a correct mathematical statement which, when translated into English, said in effect: "This statement cannot be proved."

In other words, if we go for consistency, it's possible that we've got truth, but we can't be sure. If we do manage to prove our proposition— i.e., if we could bring it to some sort of apparent self-completion—our proposition could not be true (i.e., consistent—free of paradox). Proof can appear only in a small frame, not in Totality.

We'll return to this sort of paradox later, but it's worth noting that such a great disturbance in mathematics, made by honest effort, was enough to make many mathematicians (and even some scientists) abandon the idea that they were about the business of uncovering Truth. But they abandoned the pursuit of Truth only because they discovered that Truth cannot be conceived as an object of mind.

The odd thing is that we can still *see* Truth. We just can't form an image of it in our minds. Mathematical physicist Roger Penrose wrote that, "mathematical truth is **not** something that we ascertain merely by use of an algorithm."[3] In other words, we cannot get to Truth by rules and regulation. Truth is **directly** present in the mind and needs no mediation. We either *see* It immediately, or we miss It entirely. This "missing It" is ignorance—our common habit of fixing upon concepts, on what can be formulated, born, created, modeled and held by the mind, rather than upon what is directly perceived.

Our predicament is not unlike our situation with the ambiguous Figure 1–1 on page 19. If it's not obvious to you what's going on in this figure, then you don't really *see* what appears. Rather, you're locked on to what you commonly see, a concept.

This ability we have to cast whatever appears before the senses into concept is highly developed in most of us. We do it with such rapidity and subtlety that even in this simple example we're likely to fool ourselves—at least momentarily. And this is especially true if we're very "intelligent"—i.e., quick to conceptualize, quick to "get it."

I'll give another example. Get out a picture of yourself as an infant. Is that you in your baby picture? If you say yes, go look in a mirror. You look quite different from the person in the picture—so in what way are you and the pictured person the same? If you say no, then who in the world **is** in the picture? Consider how difficult it is to arrive at a correct answer to the simple question, "Is that you?"

As may now be becoming evident in contemplating even such a simple, straightforward question, Truth—the way things actually are—cannot appear as an **object**, a concept, to the mind. And if we cannot describe or model or conceive of how things actually are, it is little wonder that beneath it all we are forever unsure of what we do believe—and, as a result, we are afraid.

We are particularly afraid when it comes to the most basic assumptions of our daily lives—the ones we don't ever allow to rise into conscious awareness, for fear that we will make no sense of common sense. I'm speaking of the unspoken axioms that we operate out of without ever giving thought to them—beliefs such as the law of identity, which says that things are what they are. We are afraid—even if only subliminally—because our beliefs, our concepts, which we commonly rely upon, give us the sense of solidity beneath our feet. But as we have seen, belief is necessarily conjoined with uncertainty; and if our uncertainty becomes great enough (through our relying evermore upon what we already believe), then our fear also becomes great. And when our fear becomes great we're all too often filled with zeal—or arrogance.

It's the fearful ones, the zealots, the earnest believers, the ones who "know" they are right and good and just and superior, who most strongly react to the uncertainty of what they believe. When we're aware of our ignorance, on the other hand, and, therefore hold no solid idea of Truth, we have nothing to cram down the minds of others. We can afford to be open and tolerant.

Being open and tolerant, we can then afford to *just see* Truth. At such a moment there's nothing to assert—and there's nothing in which to believe.

BELIEF IMPLIES DOUBT

The American philosopher C. S. Peirce noted that our commonsense beliefs are more like "belief-habits" that we do not generally scrutinize or question. For this reason they are also, as Peirce put it, "doubt-resistant." Though he maintained that most of our commonsense views would not hold through time, Peirce felt that their doubt-resistance was due to a built-in logical feature that left them vague. As an example, he cited the common belief in the "Order of Nature," observing that any attempt to give clear definition to this belief draws it into dispute. But, on the other hand, says Peirce, "who can think that there is **no** order in nature?" It seems like common sense.

Common sense is one of those entities that everyone understands but that becomes ungraspable when attempting to define it. G. E. Moore outlined much of what most of us would probably accept as common sense in his book *Some Main Problems in Philosophy*. Moore listed a number of commonly held notions, such as (a) there are great multitudes of things in the world; (b) these things are of at least two different kinds—material things and conscious acts; and (c) these things and conscious acts are subject to change (including our very notion of what common sense is).

In the rest of this book, I will primarily use the term "common sense" to mean our tacit belief in a self. This notion underlies all other ideas of common sense, for if we look carefully at Moore's (or anyone else's) definition, we find the tacit assumption of self (and the assumption of a world external to self) lurking just beneath the surface. As we shall see, these hidden assumptions are gateways to paradox.

Our commonsense beliefs are never too far from paradox. Hence our reluctance to scrutinize them, for encountering paradox would leave us with the impression that we've taken a wrong turn. And if our commonsense beliefs are wrong—if our security blanket were suddenly seen not to offer security—then what?

Most of us take steps not to get to this point. Our fear of losing security is so great that, long before we come upon paradox, we're likely to have already turned back.

It would be even more disturbing to discover what we may already suspect—that meeting with paradox when pursuing our commonsense impressions is **not** equivalent to having taken a wrong turn. Such a revelation may be enough to topple even our most deeply held belief-habits upon which we've built all that we hold most dear.

Our confusion of what is relative with Absolute occurs at such a profound level that we find it most difficult to detect our mistake—even after it is pointed out. Writer Jeremy Campbell, in *The Improbable Machine*, points out that certain unconscious cognitive processes are so important that they are "hard-wired into the brain." As a result, we often encounter "certain optical illusions [that can] still deceive the eye, even when the conscious reasoning mind knows they are illusions and tries to correct the distortions."[4] (See Figure 1–2 for an example of such an illusion.) This "hard-wired" mechanism is great for forming concepts, which give us the impression we've got a handle on what we observe. What we tend not to notice, however, is that when the conscious, reasoning mind—the conceptualizing mind—intercedes to give us this impression, we're actually further removed, as it were, from what's going on. Yet in our confusion we feel even greater confidence that **"now** I've got it!" But this "knowledge," of course, is merely another concept. It is not genuine, True Knowledge.

If we approach Truth while maintaining our commonsense view, we necessarily encounter paradox. And because of the doubt-resistance of our commonsense view, we back off in order to maintain our commonsense view. We back off out of fear that we may lose our sense of identity.

Out of such extreme fear we tighten the lid on our sense of self. Thus it doesn't strike us that our commonsense view could be faulty. But even when our commonsense view does become suspect, we are so unfamiliar with its underpinnings that we fail to *see* that our problems stem from our habitual belief in an illusion: the illusion of a self, an "I." As we shall see, under close and concentrated attention, no such "I" or self can be found—just as no definable "Order of Nature" can be found.

If we're interested in Truth, however—whether we are scientists, theologians, unschooled or anything else—we need to realize that to assume that what we think or believe actually **is** Truth is a very dubious proposition. When it comes to Truth, our beliefs put us on very shaky ground, because as soon as we latch onto an idea—a belief—the mind is immediately uncertain. Even conceptualized sensory data (as opposed to "raw," pure, perceived sensory data), which science must ultimately rely upon, create great doubt for us when we unquestioningly believe these data to be Truth, or indicative of Truth.

For example, we all know from experience, though we may not always appreciate to what degree, that problems occur when we try to distinguish appearance from Reality. Bertrand Russell called it "one of the distinctions that causes the most trouble in philosophy."[5] We inherited an awareness of this problem from the Greeks—particularly from the Skeptics—and it remains a large problem for us moderns. As astronomer John Barrow put it in *The World Within the World*, scientists "study only appearances...appearances are **our** ultimate reality."[6]

Thus, it would seem that we have no guarantee that what we study is actually Real in any ultimate sense. In other words, if the reality we study is merely relative, paradoxes must necessarily abound, since none of the answers we get will ever be found to be ultimately True. This is indeed just what bare attention to our stories always bears out, as we'll see in the next two chapters.

But Absolute Reality cannot be like this. It cannot be **merely** relative. Therefore it would seem that we cannot study what is ultimately Real, Abiding and True. We cannot place the deep-set stone beneath our feet.

Once this skeptical view dawns upon us, our commonsense reaction is to think that we might as well forget about Reality, since the relative is all we're capable of grasping. It would seem, then, that the question of whether or not there is an Absolute Reality is without meaning, because we can't get a handle on it, can't put it into concept.

But our confusion in trying to distinguish appearance from Reality is precisely the sort of thinking that results from mistaking the relative for Absolute. What if, instead of clinging yet again to what we happen to believe, we might actually find a way to study Absolute Reality? Is it possible that there's such a method of study that we commonly overlook?

THE ILLUSION PROBLEM

One way to consider this question is to ask ourselves why we commonly insist that there **is** a difference between what appears and what is Real.

Figure 1–2. The Café-Wall Illusion

We've all been tricked by appearances from time to time—or so we think. But are we actually tricked by what appears, or are we merely tricked by what we **believe** appears? In other words, are we commonly tricked by our **direct experience** of Reality, or are we tricked by **believing** our indirect **story** of Reality? Do we react to direct experience (perception), or to our idea (conception) of what is experienced?

For example, if we consider the array of squares in Figure 1–2, we might notice that the lines that define the horizontal tiers of squares appear to be tilted to the right or left in alternating layers, even though the horizontal lines themselves are parallel. We can see that the lines are parallel by viewing the array from the side, from an angle slightly above the plane of the page. But again, if we view the figure straight on, even when we "know" the lines are parallel ("in reality," as we say), they still **appear** to be tilted.

It does indeed seem that we can detect a difference between appearance and Reality. In this case, at least, we surely **know** which is Reality, because we've "analyzed" (i.e., conceptualized and studied) it. At least, that's what we think.

But the question still remains: in the case of Reality, what barrier, what division, what separateness indicates that there **is** a difference between Reality and appearance? This question, seemingly trivial if not downright silly at first, will become considerably less absurd the further we pursue it.

This question remains because we cannot actually analyze Reality. It won't go into parts. We can't hold It in concept without also holding contradiction.

Let's consider how ubiquitous such examples of "illusions" are in our commonsense, everyday experience. First of all, it's not just our sense of vision that plays such tricks on us. All our senses are equally liable to lead us into deception. By placing our warmed right hand and our chilled left hand in a single tub of lukewarm water, we experience water that feels both cold and hot at once. Our sense of smell or taste can be equally misled, as in the old parlor trick of giving an unwary and blindfolded person a slice of apple while holding a slice of pear beneath their nose. And a sound of constant pitch may appear to change in pitch if it occurs among other intonations of changing pitch—an effect often exploited by composers.

Richard Wagner, in fact, makes use of this illusion on a grand scale in his *Prelude and Liebestod to Tristan and Isolde,* where, by modulating through various tonal centers, he creates in our mind an impression that the whole, massive structure of the music itself is continuously rising. It's a stunning illusion, for the mood remains dark, heavy, dense— yet everything rises. Like a gothic cathedral whose delicate spires and upward-thrusting lines can cause dense and weighty stone to breathe of lightness and ascendancy, so Wagner obtains for us in sound a sense of lightness, a sense of rising, even within the grave and somber mood of the *Liebestod.*

I mention this particular illusion because it reveals the rich backdrop to human experience that we'll be encountering again and again in various ways throughout this book—namely, that there appears a fundamental duality, a subtle opposition that appears to reside within any single entity at any given place and in any single moment. Encounters with such fundamental duality are woven throughout all of human experience. Though such experience is common, it is just as common that we quickly dismiss it, if indeed we allow it to register at all—for even though it appears everywhere, it is inconceivable and unacceptable to our commonsense minds.

EYE, EAR, NOSE AND MIND

We are easily deceived by our senses, yet our confused reading of our perceptions of Reality is not limited strictly to things "out there," beyond the windows of our physical senses. Even our very thoughts—the pure, private and seemingly self-created objects of mind itself—also deceive. What is more, our thoughts do not remain strictly within a private mental sphere, but commingle with the "outer world" of "objective reality" in such a way that the mind, even as it acts to shape that world, is being shaped **by** that world.

A friend once told me of an experience he had as a child. He was late in getting to his lunch, and he found his glass of milk and his bowl of soup (which had by then both become room temperature) to have turned "warm" and "cold" respectively. This illusion was clearly based on his expectations, on his ideas of what things are and how they ought to be.

Let me give another example. Here our expectations, forged from ideas gleaned from a lifetime of experience with the world, shape the very registering of what we see.

Let's look at a line drawn on paper. Given the proper closed curve, it clearly forms what appears as an oval:

Yet if placed within a certain context, the same oval shape may suddenly appear as a circle:

Figure 1–3

We easily take ovals for circles. We do it all the time. In spite of this, however, it took artists a while to find the trick to creating veridical impressions of common three-dimensional objects (such as circular tables and coffee cups) on two-dimensional surfaces. When they drew actual circles to represent circular shapes, they clearly looked wrong. Eventually artists learned to stop drawing things as they "knew" them to be. They learned to draw what they saw, rather than what they thought they saw.

And so, in the interplay between the inner world of thought and the outer world of the body in its environment, we find an even greater possibility that we do not grasp what is Real or True.

But the problem doesn't end even here; it is deeply embedded in the interplay of sense and mind. For example, a delirious man may see an illusory pink elephant standing behind the real hedge in the backyard. What makes such an illusion difficult to parse out from Reality is that the illusion (the elephant) is found to be **continuous** with what we otherwise call reality (the hedge). We can never be sure, even by public experience, just where deception stops and reality begins, for agreement is no guarantee that we have not all agreed to be deceived. As long as we **believe** there remains some object "out there," that object—or more accurately, our beliefs and concepts regarding that object—lead us inevitably to a degree of doubt.

To realize Truth, however, it is necessarily to be free of doubt. But, contrary to what we may have thought, getting rid of doubt doesn't mean we must have something to believe in (as if by believing in something we could ever free ourselves of doubt). Rather, to get rid of doubt, we need to *see* clearly. *Just see*, that's all.

In answer to the question I posed earlier (page 27), there is indeed a way of study that we commonly overlook. It's *just seeing*—that is, being fully awake. *Seeing* without any mental bias—without concepts, beliefs, preconceptions, presumptions or even expectations.

Just seeing, however, is an extremely subtle discipline that few of us bother to cultivate. To the extent that we do not *see*, however, we are dead in a fundamental way, and doubt is invariably in the picture.

Even superficially, it's not too difficult to realize how not seeing is an invitation to doubt. Suppose you're working on a mathematical problem and you don't know how to proceed. Let's say that your difficulty is in multiplying two negative numbers, which you don't know how to do. Now suppose someone were to tell you that you could get the correct answer by simply multiplying the two numbers as if they were positive. You're going to have some doubt that this process will yield the correct answer. Just saying, "I believe you" will not dispel this doubt—especially if you have invested much in obtaining the correct answer. Until you actually see and understand for yourself how the process of multiplying two negative numbers works, and until you use that process yourself to obtain an answer, doubt will fester deep in your mind.

In our daily lives we function all the time with just this sort of doubt, without really noticing how unsure we are (or ought to be) about things.[7] But with honest reflection, we can find Great Doubt lying beneath the ground on which we stand.

BELIEF CANNOT REVEAL TRUTH

Belief cannot be relied upon to reveal Truth, because no matter what we believe, we're always left with doubt in our mind. Truth, on the other hand, being Absolute, must strike us with certainty.

To put it another way, Truth can only be *known* directly, without any form of mediation (such as a concept or belief). It's precisely because the objects of our beliefs must be mediated that our beliefs are, of necessity, associated with uncertainty.

The question of belief arises only when there **isn't** any *knowing*—that is, only when we're not sure. For example, if I were to approach you with a closed hand and tell you that I have a marble in my hand, you may or may not have any reason to doubt me. Either way, however, so long as I keep my hand closed, your mind must remain in a state of uncertainty about my possession of a marble. Under these circumstances, it would make perfectly good sense for you to refer to the presence of a marble in my hand as a matter of belief. It would **not** make sense, however, to speak of this as *known*. You may believe that I hold a marble or you may not; but, Reality being what It is (Absolute and not dependent upon what you think), what you happen to believe has little to do with it. So long as my hand remains closed, you could say that you believe I'm hold-ing a marble—or that I'm **not** holding one—but you would have to tack on, "However, I'm not really sure."

Actually, it's not at all unusual to couple a statement of belief with a confession of uncertainty. No one is likely to ask, "How is it that you can say that you're not sure after stating that you believe it?" In fact, often when we speak of a belief, we phrase it as, "Well, I'm not really sure, but I believe it's like this…." The believing mind always exists in a state of doubt.

Rather than gathering even more things and ideas to believe in, the way to get rid of doubt is to *just see*—clearly, directly. Ridding ourselves of the Great Doubt requires direct, deep, penetrating insight into the nature of Absolute Reality.

GREAT DOUBT

Great Doubt is not like ordinary doubt. An ordinary doubt would arise in your mind if someone were to tell you that they could speak twenty languages, or that they saw unicorns romping in the hills behind your house.

Great Doubt is more like this: hold your hand in front of you. If you doubt that you are looking at your own hand at this moment, that would be Great Doubt, for you'd be doubting the immediate *this* before you now.

Great Doubt means to doubt even the very bottom of what you've al-ways accepted as True without question. If you can doubt *this*—imme-diacy—that's Great Doubt. This isn't a matter of **saying**, "Yes, I doubt what I see." You can't doubt by simply saying you doubt, any more than you can believe by saying you believe. You can't **choose** to doubt, for doubt is merely the flip side of belief.

For example, you didn't choose to doubt the person who claimed they saw unicorns chasing about in the hills. You simply doubted them right off. In like manner, Great Doubt is when you doubt *this* straight off, but what you Doubt in this case is the immediate object of consciousness—*this*. When you look at your hand and experience doubt—immediate, silent, unnamable doubt—that's Great Doubt.

Great Doubt can come at moments when we're overwhelmed by a profound sense of "this cannot be, this cannot be!" Such a feeling may overtake us at the moment we realize that we're about to die.

I had this experience years ago when, after discovering I had cancer, a physician told me I probably wouldn't live another three months. Getting such news, and getting it suddenly, immediately thrust me into a state of shock and fear.

But the circumstances surrounding Great Doubt need not always be so dramatic. I saw this same state of mind sweep over someone at the moment she realized that her car had been stolen. In fact, "This cannot be, this cannot be," were her precise words at the moment of her realization.

These moments that come to us with such intensity are very important, because through such intensity we can most easily awaken to what is actually going on. That is, we can awaken to *this*—to immediacy. We can awaken to experience prior to concepts.

This is why a not uncommon time to wake up is just before death. The "it cannot be!" overtakes us in that moment.

But we do not need to wait until the last biting moment to be smitten with Reality. Just as we can place ourselves in a position that fosters particular beliefs, so, too, we can position ourselves to experience Great Doubt. Only then is it possible to *see* through Great Doubt, to go beyond paradox and confusion, and to obtain genuine Certitude.

DOUBTING THE GROUND OF COMMON SENSE

Going beyond paradox and confusion is, in one sense, not easy. Typically, it requires some training and experience in concentration, and a lot of practice. There are many traps and snares along the way—all conceptual. But if you observe the mind's work (i.e., watch its reflexive action in attending to its objects), you can acquire mental habits and attitudes that avoid these traps.

American philosopher and logician W.V.O. Quine highlights one such set of traps. He points out that experience never forces us to reject any particular belief, for we can always modify parts of our belief system to accommodate new awarenesses, or new interpretations of experience. Thus only when we stop confusing what we believe with what we *see* are we no longer in danger of the traps Quine points to.

But what we're concerned with here is getting beyond Great Doubt—extricating ourselves from some sort of pseudo-Absolute Belief. We first must remove our very ground of belief until nothing remains except Great Doubt. Then we must—somehow—eradicate Great Doubt as well, until not a shred remains.

This state of mind is possible, though rare.

We must doubt and doubt again—doubt to the very bottom, to the ground, and then doubt the ground itself.

How do we come to doubt the very things we once believed? Indeed, how do we come to doubt what we've never even thought to question? How do we turn the coin over? How do we go from believing in things (our concepts, our mental objects, and our stories) to **truly** doubting them, and thus sending belief into the shadows?

How do we doubt the very ground of common sense?

The best way is to carefully attend to what's actually going on, and to notice that our stories never fully explain it. After long slumbering in habitual, confined thought, the mind comes to a new awareness and is transformed.

This careful attending to what's happening, though uncommon, is by no means beyond the reach of ordinary people. A wonderful example of it occurs in the writings of the explorer Amerigo Vespucci. Vespucci visited the "antipodes," a word that means the other side of the Earth—literally, a place where people's feet are opposite. Suddenly confronted with a world more vast and varied than his belief system could account for, Vespucci lost his innocence. We can almost feel the shift in his mind as his belief was transformed into doubt:

> What should I tell of the multitude of wild animals, the abundance of pumas, of panthers, of wild cats, not like those of Spain, but of the antipodes; of so many wolves, red deer, monkeys, and felines, marmosets of many kinds, and many large snakes?[8]

He was led to the then-heretical conclusion that "so many species could not have entered Noah's ark."

Vespucci did not fall into the trap that Quine warned us of: he did not frantically attempt to modify his old, familiar explanations (by concluding, for example, that Noah's ark must have been miles long). Suddenly he became too informed to do that. He was unable to go back to his former naïve belief in the Biblical account of The Flood.

This is how we come to doubt. Our eyes must remain open long enough that we may be overwhelmed by a new experience, a new awareness, that shatters our habitual thought and our familiar stories. Einstein's mind was overwhelmed in just this way by physical evidence of an expanding universe.

We should always be willing to take another look at what we believe and begin to doubt it. As we shall see, it's always possible to doubt **any** concept or belief.

We should look and doubt even beyond the profound skepticism of Pyrrho and Sextus Empiricus. They and the other Greek Skeptics asserted that we live in a dual world of appearance and Reality. Since all we can *know* are the appearances, the Skeptics argued, and appearances seem to deceive, we therefore cannot have any guarantee of Truth. I will argue, however, that: (1) these assertions are themselves derived in concept and, hence, doubtful (Sextus Empiricus seems to have gone this far himself); (2) that it's not possible to correctly register in consciousness (i.e., conceptualize) what actually appears to perception; and (3) it's out of such ignorance that we posit something that does not in fact actually appear in direct experience—a "self."

We should go beyond the Skeptics and doubt until we no longer look to our ideas of things at all, but instead *just look,* and *just see.* We should doubt until we're assured we're not overlooking something, that we're not constructing something, and that we're not taking something for granted. We should doubt until we no longer hold fast to any thing or idea at all.

Most of the time we are desperate to place (or replace) a floor beneath our feet. We try to ignore belief's inevitable links to doubt. But deep beneath the surface, our mind spins in confusion and smolders in Great Doubt. It's this profound Doubt that we must get in touch with if we would find peace of mind.

Were I today to claim a belief in, say, the literal story of Noah's ark, no one would think that I possessed a charming innocence that they might otherwise grant to people of the Middle Ages. Rather, I would simply be considered ignorant. The story of the ark is a wonderful myth, and it says many valuable things about human beings, but it doesn't belong to our Age. We ought to know better than to accept it literally.

But we cannot stop there. If we pick up any other belief and run with it, we will, in time, discover that as far as Truth is concerned, we are still as ignorant as we ever were. We still have not freed ourselves from the possibility—indeed, the certainty—that confusion lies within our story.

We are free from this confusion only when we cultivate an open and inquiring frame of mind, ever on guard not to sink into insisting upon some particular belief, no matter how well "justified."[9] As we shall see, **no** belief can ever be fully justified.

If it's Truth we're after, we'll find that we cannot start with any assumptions or any concepts whatsoever.

Ignorance has no bounds. It is ever with us. Our fallen state is not the result of eating from the "tree of knowledge," but of constantly eating from the "tree of concept" without understanding the consequences.

We'd suffer less if we'd only learn to cultivate an eye that *just sees.*

ERGO SUM?

The subtlety with which we make many of our assumptions is profound indeed. In fact, there is a basic unwarranted (i.e., not found in experience) assumption that goes unidentified by nearly everyone all the time. The consequences of this assumption—this ignorance—are great.

The best example of this unwarranted assumption appears in Descartes' classic proposition, "I think, therefore I am."

Descartes wanted to get to some statement that could not be doubted. He wanted Certitude. In his day, religious authority had fallen under attack. There had been a renewed interest in the ancient skeptics, most notably Sextus Empiricus, and in the unsettling idea that all propositions could be rendered equally improbable. It was even being seriously questioned whether there could be any Knowledge at all. This was a very troubling problem in Descartes' time, and it still is.

Descartes contemplated the possibility that **all** we commonly believe might be false. For him the question was, "What do I know?" He tried to find the answer by searching through the various beliefs he felt inclined to hold and, though he was not a skeptic himself, he used skeptical methods to bring himself to doubt everything, even beliefs he had long held. He doubted all until he came upon his *cogito*—"I think"—which he regarded as a "primitive datum that the mind can recognize only when it encounters it."[10] And here he stopped, thinking that he had hit bottom.

Descartes set down the assumption that "I think" is the ground that is beyond all doubt. But even in the simple statement "I think," Descartes had already made an assumption: he assumed the existence of a self, an "I."

Once he had done this, of course, it was not too difficult for him to "prove" the inevitable conclusion ("therefore I am"), since he had already assumed an "I."

Such a tightly knit package appears as a truism to most of us—and, indeed Descartes' *cogito* does appear as a truism to many. But Descartes did not doubt enough. In saying, "I think, therefore I am," we have already assumed the "I's" existence before we begin. This merely reflects our normal way of thinking—we **all** assume a self most of the time.

But this is no proof. Just as it is not difficult to discover "God" if we begin with the foregone conclusion that "God is," so too it is not surprising that Descartes could discover "I am" after he had already posited the "I" in his thought. This isn't the unshakable proof, the indubitable ground that Descartes was seeking. If he truly questioned his existence, how could he have already assumed it?

> I noticed that while I was trying to think everything false, it must
> needs be that I, who was thinking this, was something. And ob-

serving that this truth, **I am thinking, therefore I exist [Je pense, donc je suis]** was so solid and secure that the most extravagant suppositions of the sceptics could not overthrow it, I judged that I did not scruple to accept it as the first principle of [the] philosophy that I was seeking.[11]

But where or what is this "I"? Does Descartes mean his mind is thinking? Does he mean his body is thinking? Notice how the "I" gets tossed into the picture from out of nowhere. What is doing the thinking here?

The absurdity of this assertion becomes clearer once we switch subjects. We've all used the common expression "It's raining." But would we say, "It is raining, therefore it is"? What is raining? Do we suppose there is some entity corresponding to the word "it" that is doing the raining? No, of course not.

But how does this situation differ from "I am thinking"? What is raining? Who is thinking? Where are these hidden entities? What is this "I" we keep referring to? What thing corresponds to this word? What is this "I" that is doing the thinking?

You may say, "But, of course, when we use the expression 'it is raining,' there's nothing out there that corresponds to the word 'it'—it's just that we cannot construct a proper sentence in English unless it has a subject. And so, by convention, we insert one. But this is obviously not the case with the 'I' in 'I am thinking.'"

Oh? Then what **is** the "I" referring to? Where is it located? What are its properties? The more we try to grasp what "I" is, the more it slips away from us. As Ambrose Bierce put it, "I think I think, therefore, I think I am." We've assumed self at the end of an endless regression, and we can't get to it. "I" seems to refer to something we tacitly assume is there, but that we can't seem to find. The "I" is deeply, profoundly, and quietly assumed, but it's assumed without justification.

You'll recall from earlier in this chapter how difficult it is to answer a simple question such as, "Is that you in your baby picture?" We can't find anything in experience that clearly corresponds to the word "you." That is why we find it so difficult to answer this question.

So, how does "I am thinking, therefore I am" differ from "it is raining, therefore it is"?

"I" is a most difficult thing to come to doubt—but we must go beyond Descartes and doubt it, for it is in fact no more than a mere construct, a concept, a belief. It is nowhere to be found within direct experience (i.e., through perception). **We do not find anything in our experience that corresponds to that word, "I."**

We'll be looking at this matter in detail later—but, before we go any further, let's consider how Descartes might have constructed his "proof" in a manner that better reflects actual experience—i.e., that reflects perception rather than conception.

Note that, short of its being demanded by the conventions of language, the "I" is absent from actual experience. What Descartes was directly aware of was thought, not "I." Whichever way we might put it—"thought," "cognizance," "awareness," "mind," "consciousness"—these words directly refer to immediate experience. The word "I" does not.

The problem with Descartes' *cogito* is that it posits a self before it acknowledges thought. But since it is thought and not "I" that is directly experienced, then Descartes might have said, "Thought, therefore I am." But such a proposition is clearly absurd, for now it is plain to see how the "I" just pops into the picture out of nowhere. The insertion of "I" follows neither logically nor experientially from the first statement.

To get his statement more in line with direct experience, Descartes might have said "cogitatio ergo esse"—"Thought, therefore being." Or, "Thought, therefore existence." **Thought, therefore something (without naming it) is.** *Something's* going on, in other words.[12]

We must come to doubt the commonsense idea that we experience an "I" antecedent to, separate from, and distinct from what is "not-I." We must come to doubt the explanation that there can be a self (literally a "not-other") that is separate from an other.

How we actually do experience (i.e., perceive rather than conceive) the world will be discussed in Chapter 4. For now, it's enough to say that neither "self-separate-from-other" nor "self-not-separate-from-other" are correct reflections of actual experience.

TRUTH CANNOT BE BELIEVED

Truth must be *known*. It cannot be simply believed. Belief, holding an opinion, is necessarily conceptual. Truth, on the other hand, being Absolute, cannot be conceptualized. Truth is not what can be believed, for It can't be formulated in a phrase and held by the mind. It cannot be "understood" as such. Yet It can be *seen* and *known*.

You'll recall the example I gave earlier in which I held out my closed fist and told you there was a marble in it. Now, you may or may not believe I'm actually holding a marble; but when I open my hand, it suddenly becomes meaningless to continue speaking of belief, for now the presence (or absence) of the marble has become known. Like this, Truth, once actually *seen*, is already *known* **directly**. At that moment all uncertainty clears up.

Now you may very well protest and say, "But surely knowledge is also conceptual." And, indeed, what we **commonly** call "knowledge" surely is, for it's actually belief. And though philosophers since Plato have rejected the idea that knowledge and belief are one and the same—mostly on the grounds that knowledge, unlike belief, provides Certitude—the fact is, conceptual knowledge **never** provides Certitude, as we shall see.

Unlike knowledge, Knowledge or Certitude can come only with direct perception, never with conception. What we commonly assume of knowledge—that it is conceptual—is, in fact, a necessary characteristic of belief, but not of True Knowledge.

In these subtle ways we confuse Knowledge (perception) with belief (conception). As a result of this confusion, contradictions necessarily occur in all our stories, and in all our attempts to account for our experience.

The real distinction, then, is not between belief and knowledge but between belief (i.e., what can be doubted) and Certitude. Whether we call our concepts beliefs or knowledge, they are necessarily linked to doubt. True Knowledge, or Certitude, is pure, objectless Awareness. It isn't linked to doubt, or concept, or belief.

What most distinguishes Certitude from belief is that belief always involves an object, while Certitude, since it must necessarily be immediate, does not. Another way of putting it is that belief yields paradox, while direct Knowledge does not.

It's quite possible to believe all sorts of things that are not true, but we can only have actual Knowledge of Absolute Truth. Short of this, our "knowledge" can be nothing more than belief.

But how might we establish Real Knowledge? What would be necessary in order that we may be certain that what we perceive is Real and True? We might well wonder if complete certainty could ever be obtained, since at any point we may unwittingly be mistaken about what is being perceived. If we can only have Certain Knowledge of Absolute Truth, and if it is impossible to conceive of objects that are Absolute, it would follow—seemingly, anyway—that we can have no idea of what is True.

To declare that there is no possibility for Certitude, however, is already contradictory. If there truly were **no** possibility for Certitude, then we have attained Certitude about the impossibility of Certitude! We thus become stuck in a vicious circle, because we assert what we deny.

Part of our problem of why we don't ordinarily get to True Knowledge is that it must **be** Certitude—and we reject Certitude. Instead, we insist on "getting it"—having Knowledge—by putting conceptual handles on It. And then we get frustrated, because It won't go into concept. In our frustration, we may begin to lower our standards for Certitude. Indeed, G.E. Moore, in a move somewhat reminiscent of the Academic Skeptics, argued that our problem stems from the fact that we've taken Certitude to mean 100% certainty, which is what everyone thought it meant before he said it didn't. But accepting anything less than True Knowledge for Certitude simply means once again to replace Certitude with belief (and, thus, with a degree of doubt). The same problems immediately arise, and we are back to where we started.

Yet only Certitude can still the hollow ache of the heart. No substitutes will satisfy.

All of these intractable problems (and, as the centuries since the ancient Skeptics have come and gone, they've only become more intractable) are brought on by conceptual thought. Our problems stem from the fact that we tend not to appreciate direct perception, but instead build upon false assumptions.

So let's turn our attention to these false assumptions—what we unwittingly take for Knowledge—and see how, like belief, conceptual knowledge is never without Great Doubt.

two

⧽(KNOWLEDGE)⧼

These our actors,
As I foretold you, were all spirits, and
Are melted into air, into thin air;
And, like the baseless fabric of this vision,
The cloud-capp'd towers, the gorgeous palaces,
The solemn temples, the great globe itself,
Yea, all which it inherit, shall dissolve
And, like this insubstantial pageant faded,
Leave not a rack behind.
— FROM *THE TEMPEST*,
BY WILLIAM SHAKESPEARE

In a substantialist view, the universe will be un-
born, non-ceased, remaining immutable and
devoid of variegated states.
— NAGARJUNA

GOING TO ZERO

Science is the religion of our time. Much of what we assume of Reality, our *Weltanschauung*, has been shaped by the general view that is embraced by most scientists today. Often called "realism," it is a "substantialist" view. This scientific realism, according to physicist James Cushing, "requires roughly at a minimum that our scientific theories are to be taken as giving us literally true descriptions of the world."[1] Yet we have already seen that all conceptual thought—all theory, concept and belief—leads inevitably to contradictions. Indeed, some scientists themselves view scientific realism as suspect. Physicist Asher Peres, for example, reaches the conclusion that "any attempt to inject realism in physical theory is bound to lead to inconsistencies."[2]

In this chapter we'll examine what we believe we know. We'll question the view of scientific realism and test its foundations. We'll find that it's not merely this system that is without foundation, but **all** belief systems. In the process, we shall gain a radically different perspective on some of our most profound problems of Reality.

We'll see that the problem isn't merely some logical limitation or inbuilt defect in our methods. We will see that we do not find a conceptual ground to experience at all, for the same reason Flat-Earthers do not find an edge to the Earth: because there is none. We'll see that what "ground" we **can** find is utterly nonconceptual.

Scientists, on the whole, make certain basic assumptions, all of which turn out to be unwarranted and unfounded. Like Descartes, they begin at step one instead of at zero. This chapter, however, and Chapter 3, are about going to zero. We'll mark several trails, all starting with a few commonsense assumptions, and we'll follow them all to zero. As we'll discover, **any** starting point other than zero (i.e., any belief system) will lead us to zero, if only we would attend fully to perception alone.

Most of our effort, our thought, our habits, our desires, our culture and our education is designed to suspend us in conceptual thought. In this chapter and the one that follows, we'll attempt to break through conceptual thought.

SCIENCE AS A SYSTEM OF BELIEF

Scientists are in the business of knowing. As physicist Johann Rafelski put it, "Science is about knowing. It's not about believing."[3] Science, we say, is not a belief system, but rather a methodical search for Knowledge. Science is a way of going about the world in search of what can be established as "justifiably true"—which is how contemporary philosophers define knowledge.

Our modern definition of knowledge, however, as a "justified, true belief" was dealt a serious blow in 1963 when Edmund Gettier showed that one can have a justified, true belief and yet not know what one believes. His argument runs like this: say a man believes there is a sheep in a field, but it is actually a dog that he's mistaken for a sheep. Yet, as it turns out, there actually **is** a sheep in the field, but it remains unseen by the man. The three criteria for knowledge (belief, justification, and truth) appear to have been met, yet we cannot say that this person actually knows there is a sheep in the field, since his "knowledge" is based on having mistaken a dog for a sheep.

Since the arrival of the "Gettier problem," as it has been called, others have put forth new ideas of adding yet a fourth criterion—e.g., that knowledge is a "nondefective" or an "indefeasible," justified, true belief. But, as we'll soon see, adding this fourth criterion does not get us any closer to Knowledge or Certitude. Indeed, piling up criteria turns out to be utterly futile.

As it's commonly practiced, science is our attempt to arrive at concepts that yield greater and greater doubt-resistance—that is, concepts that come with stronger and stronger justification. Science gets in there and examines the world, carefully and in great detail. No theorizing is taken on faith; every theory is put to the test. The irrational beliefs of scientists, and their biased attachments to pet theories and projects, shouldn't have many deleterious effects in the end, for everything is open to peer review. The effects of human weakness and folly get ironed out over time. In short, science is honest work done in the open. Anyone can repeat an experiment and verify or reject what the experiment purported to prove.

Clearly, science has been humanity's one great attempt to get to the bottom of things. And so, we tell ourselves, we can put our faith in science. The conclusions—concepts—we arrive at through the scientific method come only after slow, hard, thorough research, yet even then, we maintain that all is subject to being overthrown by further research and information. What more can we do than this?

Actually, science is the predominant belief system today. Even those of us who possess very little knowledge in science still treat it somewhat as a religion. As a society we put our faith in and make use of the "miracles" of science. For the most part, we believe that the beliefs scientists hold about the universe are indeed justified and true. And it's our scientists to whom we typically turn for answers, explanations, and wisdom, much as people in earlier cultures turned to shamans, village elders, and medicine men.

At the same time, however, we are aware that our scientific beliefs are subject to change and modifications as the result of future research and discoveries. This is rather curious. We seem to be willing to accept what science tells us—and equally willing to accept that what we have so easily accepted may turn out to be false! In short, we accept science **only** as a belief system, never as a source of Truth, Knowledge or Certitude. (Of course, it's rare that a whole platform in scientific theory is dismantled. Usually only a few planks get replaced or removed or turned around.)

We've managed to convince ourselves to accept a system that can yield only a "maybe" at its best. As it was put in the *Skeptical Inquirer* by Lys Ann Shore,

> The quest for absolute certainty must be recognized as alien to the scientific attitude, since scientific knowledge is fallible, tentative, and open to revision and modification.[4]

We no longer believe our science **is** about the search for Truth. And so, because relative knowledge is all that science (or any belief system) is capable of dealing with, relative knowledge is all that science ever finds. And it's all we've come to expect is possible. Though science astounds us in how precisely it has allowed us to define and manipulate the physical world, when it comes to enlightening humankind on ultimate Truth and Reality, it fails. In the end, science is not capable of providing Certitude. And this is fine, as long as we don't conclude that Certitude is therefore impossible.

Even so, science is in the business of acquiring knowledge—that is, justified beliefs. Yet science **rests** upon an **unfounded** belief. Science, not by necessity but by common assent, rests upon the enormous commonsense assumption that an external world is Really "out there." As we shall see, we cannot assume this without rushing headlong into paradox.

THE RELIGION OF SCIENCE

In his book *The World Within the World*, astronomer John Barrow observed that "the practice of science…rests upon a number of presuppositions about the nature of reality. We usually take them for granted." He then quoted the Hungarian scientist and philosopher Michael Polanyi:

> the metaphysical presuppositions of science…are never explicitly defended or even considered by themselves by the inquiring scientist. They arise as aspects of the **given** activity of enquiry, as its structurally implicit presuppositions, not as consciously held philosophical axioms preceding it. They are transcendental preconditions of methodological thinking, not explicit objects of such thinking; we think **with** them and not **of** them.

Barrow then lists nine of these presuppositions:

1. There exists an external world which is external to our minds, and which is the unique source of all our sensations.
2. This external world is ultimately rational. 'A' and 'not A' cannot be true simultaneously.
3. The world can be analyzed locally without destroying its essential structure.

4. The elementary entities do not possess what we call free will.
5. The separation of events from our perception of them is a harmless simplification.
6. Nature possesses regularities, and these are predictable in some sense.
7. Space and time exist.
8. The world can be described by mathematics.
9. These presuppositions hold in an identical fashion everywhere and everywhen.[5]

For good measure, I'll add a tenth: "A thing is what it is." This, of course, is the law of identity: a thing is identical with itself and implies itself.

Barrow says that the presuppositions of science "enable us to proceed most effectively from simple experience of the world to knowledge of the world."[6] But this is precisely how we confuse belief with Knowledge! As Barrow's statement reveals, we have already missed what bare attention provides as base experience, and replaced it with a set of beliefs (what he calls presumptions).[7] It's Descartes' fundamental error all over again.

Most scientists, and indeed most people, believe there's a great deal behind these propositions. As science writer Martin Gardner puts it:

> The hypothesis that there is an external world, not dependent on human mind, made of **something**, is so obviously useful and so strongly confirmed by experience down through the ages that we can say without exaggerating that it is better confirmed than any other empirical hypothesis.[8]

It is not difficult to find others who agree. Mathematician Morris Kline, for example, has written that, despite the denials, qualifications, and reservations of certain philosophers:

> physicists and mathematicians do believe that there is an external world. They would argue that even if all human beings were

suddenly wiped out, the external world or physical world would continue to exist. When a tree crashes to the ground in a forest, a sound is created even if no one is there to hear it. We have five senses—sight, hearing, touch, taste, and smell—and each of these constantly receives messages from this external world. Whether or not our sensations are reliable, we do receive them from some external source.[9]

The observation is repeated ad nauseam: mathematician John Casti referred to a straw poll taken recently in a small university's department of physics, where ten out of the eleven members of the faculty "claimed that what they were describing with their symbols and equations was objective reality. As one of them remarked, 'Otherwise, what's the use?'"[10] Similarly, when Copernicus replaced the Earth with the sun as the center of the solar system, he believed he was offering a description of how things "really are." This has been the dominant attitude of scientists ever since. Physicist and author Nick Herbert, who has superbly and insightfully presented the bizarre realities that seem to lie behind the "phaneron" (the phenomenal world), has argued that, unlike some philosophers, but like ordinary people, "physicists cannot deny the evidence of their senses. The indubitable reality of measurement results is a solid rock on which to found an empirical science, or from which to launch speculative voyages into deep reality."[11]

I could go on citing such comments, but clearly belief in an objective reality and an external world is a central tenet of modern scientific faith.[12]

But such a belief is not a very solid rock, I'm afraid. Gardner believes that no one "except a madman or a professional metaphysician"[13] would doubt such a belief. But I would argue that an empiricist fully attending to what is provided **by perception alone** would doubt it, and I would have us doubt it here.

Oddly enough, after declaring that only a madman or a metaphysician would doubt an external world, Gardner adds that this hypothesis says "nothing about the essential nature of the external world; only that

something lurks behind the phaneron to preserve its complex regularities."[14] But what is this lurking **something** and why is it there at all? Or more appropriately, what is it doing "out there"? Like Bertrand Russell, who said that for him the great mystery is why there is something as opposed to nothing, we **do** feel **something's** "out there."

For Bishop George Berkeley the **something** "out there" is the mind of God. For materialists ("substantialists") the **something** is an objective reality. But to one who attends fully to what is given in experience and not to thought constructs—in other words, to a pure empiricist—there's no ground to support the notion that there's a regulating "behind" to the phaneron. There's not even ground to support the idea that there's any substance to the phaneron's "front"! I'll say more about this shortly.

The power and validity of science would seem to arise from the apparent fact that it relies on empiricism and indubitable mathematical deduction. But the fact is that science rests not upon any such solid ground at all, but upon presumptions that, by their very nature as presumptions, must harbor doubt, and upon deductive reasoning that must remain uncertain so long as these presumptions are rooted in the metaphysical and not the empirical.[15] Science thus rests upon nothing solid, but merely examines and assists in an endless series of furniture rearrangements in a room. Science, as it's currently practiced, will never lead us to glimpse the nature of the room itself.

Furthermore, scientists **must** believe in an external world, simply because it's the task of the scientist to measure, test and observe the world "out there" so that conclusions about reality (or at least about phenomena) may be drawn. In other words, without a belief in an external world, science itself cannot proceed—or so, at least, it would appear.

HERE IT IS, BUT WHAT IS IT?

I do not mean to argue that an external world is **not** "out there"—nor am I arguing the converse. I am simply suggesting that the existence or denial of an external world are in fact propositions we cannot make with any validity. In fact, I intend to demonstrate that the question of the existence vs. the nonexistence of an external world is meaningless—much as questions regarding the edge of the Earth have been rendered meaningless.

I'm not the first to come to this conclusion. Søren Kierkegaard, the Danish philosopher and theologian, thought that no accurate model of Reality was possible. Kierkegaard, however, felt that Reality contained a fundamental ambiguity or paradox that would forever block our vision of Truth. I am suggesting, however, that what appears as a fundamental ambiguity or paradox does **not** block our vision of Truth, but, rather, leads directly to it. Indeed, to abandon our pursuit of Truth simply because no model can be made is to give up precisely when the first glimmer of Truth is present.

Let's look at the phenomenal world. The Realness of there being anything "behind" phenomena is questionable. If we read the writings of such philosophers as Berkeley, Locke and Hume, we have to consider that what we are aware of, when we think we are observing a world "out there" apart from ourselves, is nothing more than our sensations. As Berkeley pointed out, if there were an external world, we should never be able to Know it; and if there were not, then we should have the same reasons as now to think that there is one. As we have already discussed, this observation cannot lead us to any solid ground—but it does indicate the need for us to leave our belief (in either an external world or the lack of one) suspended.

Nevertheless, **something**—phenomena, at least—is there. Something is moving our senses—or, so it seems. But what is it?

This question—"what is it?"—arises with the appearance of something—that is, with any mind object. It generally goes unnoticed, though, because we're so quick to conceptualize experience and explain it to ourselves in familiar terms. The "what is it?" aspect of experience, however, often becomes noticeable when we see something (the mind object) from an odd angle, or in dim light, or under some unusual circumstance—or when what we see is simply unfamiliar. For an instant (or, in rare cases, for several seconds) after we first make out an object in our mind, we do not know what it is, or what to call it, or how to respond to it. In that instant of awareness prior to recognition, we may feel uneasiness or even outright distress. We then struggle to re-frame bare perception into familiar terms once again, hoping to scratch the "what is it?" itch in our mind. Thus we easily buy into some definition or label ("that's a banana squash"). Once this occurs, our attention to our object, and to what is actually taking place, diminishes greatly.

We have not really adequately answered the question of "what is it?" We have merely answered the question "how do we conceive of it?" or "what do we call it?" Some deeper question remains. And if we continue to scrutinize our mind object, sooner or later the question will reappear. In fact, if we just persist in strictly observing things, however they happen to appear, the question—what is it?—invariably recurs, a persistent and troubling uncertainty. We never arrive at anything solid.

For example, if I say, "Here, in this cup, is water," you may ask, "What is water?" We could end our discussion at this point if you were to just take a drink. But as scientists we might wish to point out, "Water is hydrogen and oxygen." (This would not be an answer we could give on the basis of having drunk some, of course—that is, on the basis of direct experience. We can obtain this answer only after we have conceptualized and analyzed the water very carefully.)

Thus by using scientific methods it seems we can discover what water is "made of." With confidence we say, "What is **really** in this cup is hydrogen and oxygen, combined and transformed into this unique substance we call 'water.'"

But the questions continue. What is hydrogen? What is oxygen? And so we look again, using scientific methods, and say, "Hydrogen is an element made of atoms, each consisting of a single proton and a single electron." But still the questions remain: what are atoms? What are protons and electrons?

It seems that we've started on a never-ending regression. At no time do we ever really get to the other end of the question: "What is water?" We can name the mind object, even break it down and name its parts, but we still don't really answer the question. In the end, water (or anything else) is just like the banana squash I encountered in the farmers' market. We can discover what it is called, but we can never Really say— or Know—what "it" is. Yet, paradoxically, we can experience *what's* going on. We can drink the water.

When we look "out there" for the answer to "what is it?" we find endless regression. We can only point to some other thing (or set of things) and say, "it is this" (or "it is like this"). But try as we will, we can never gather all of what "it" is together into one place to reveal what it Truly is.

In fact, phenomenal reality **always** presents itself in human consciousness in the form of: "here it is," **and** "what is it?" I'll henceforth refer to these two aspects of phenomena as *this*, and *what*. Here's the cup *(this)*, but what is it *(what)*?

What can be more accurately thought of as pure interrogative. It's a state of mind often depicted in the comics as a question mark appearing over the head of some bewildered character. It's the fragile state of mind I had when I happened across the banana squash at the farmers' market. **It's a state of mind we will inevitably come to if we persistently analyze the phenomenal world.**

It's this *what* aspect of phenomena that our commonsense view of the world typically overlooks. (In fact, common sense **demands** that we overlook this aspect of Reality.) But it's also this very aspect that determines that science cannot reveal Absolute Truth, for science can never truly answer, "what is it?" It can only answer, "what is it called?" and, superficially, "what is it made of?"

If we try to ignore this troublesome *what* aspect and examine the presumed external world in detail, and if we go far enough in our inquiry, we'll discover that we can't get a conceptual handle on things. Rather, we'll find that the *what* aspect will appear to us in at least three ways: first, that an objective world can't be discerned from what is subjective; second, this presumed substantial, external, physical world will eventually appear devoid of all substantiality; and, finally, the only truth revealed through the study of an external world is merely relative.

Let's look at these three points more closely. You may note that, as we consider these, the distinction between mental and physical phenomena will become considerably less clear.

AN OBJECTIVE WORLD CANNOT BE
DISCERNED FROM WHAT IS SUBJECTIVE

A cup of coffee sits on my desk in front of me. From five or ten feet away, I can see the cup very clearly. I can hear and feel it as well, if I snap my finger against its rim. When I include a relatively large part of the world that surrounds the cup—the air, the light, my finger, my eyes and ears, the desk beneath the cup, the room in which the cup appears, etc.—I can discern "cup" quite easily.

But suppose that you and I, using scientific instruments, move in closer for a better look. When we do this, we quickly lose the "cup of coffee." First we see just a ceramic wall. Examining more closely, we find merely a lot of rapidly moving molecules. At this point we are no longer viewing anything that we may rightfully call a "cup." Our object has now become a collection of molecules.

Once we're in close enough to "observe" the cup's atoms, we start to notice that something very strange is happening. The atoms, which we say "make up the cup," seem to be losing many of the properties we attribute to everyday, commonsense, physical entities such as cups, clouds, planets and people. Atoms seem to have less definite positions in space, for example. They seem, rather, to be somewhat fuzzy or indeterminate.

If we get in close enough to view our object on the level of the subatomic particle, we find that these very minuscule bits of matter (can we call it matter at this point? If we cannot, then where did the matter go?) simply do not have qualities such as position, or momentum, or size, or velocity, or any number of other such physical attributes.

At this point, we have not only not answered the question "What is a cup of coffee?" but we have ended up posing several others: "What are molecules?" "What are atoms?" "What are subatomic particles?"

Furthermore, the closer we look at some of these things, the more bewildering they become. An electron's position, for example, is not something that really exists—until we look for it. Electrons have specific locations only when someone is looking, it seems. Until we looked for it the electron didn't possess anything that we would commonly call a position. On the other hand, if we look for its position and nail it down—to a general area, anyway—it seems that, by virtue of our knowing its position, we've now forfeited the possibility of knowing much about its momentum. And if we choose to look for an electron's momentum rather than its position, we would be able to measure that momentum, but we would discover that the electron doesn't seem to have a position! This is what physicists refer to as the Heisenberg uncertainty principle. It's an essential ingredient of physical reality.

This is precisely the sort of thing science finds when it takes a close look at phenomena. Without the consciousness of an observer, the stuff underlying this physical reality does not seem to exist. Only when we look for something does it appear to leap into existence—and, at the same moment, what we do not look for cannot be said to exist.

We tacitly assume that Reality only presents us with *this* (our object of consciousness). But we don't know what to make of a Reality where things are instead weirdly blended with, or take their identity from, what they are not.

No matter how we slice it, this is physical reality at close range. Subjectivity, it seems, enters the "objective" world at a very profound level.

THE INSUBSTANTIALITY OF THE PHYSICAL WORLD

The second reality we discover, when we attempt to put to rest the *what* aspect of objects through a careful study of the material world, is that substantiality disappears. When we drink coffee from a cup we naturally assume the cup is "there." We say it's "substantial." But what are we talking about? What does it mean to be substantial?

We say the cup is made of atoms, which in turn are made of subatomic particles. Yet if we take two subatomic particles—say, protons—and smash them together at extremely high speeds, we find that the two original colliding particles fly apart, **along with two new additional particles.** The two new particles didn't exist anywhere in time or in space before the collision. Physicists have done this repeatedly, with the same results every time. One physicist said it would be like smashing two watches together, but in addition to the expected wheels, springs, gears and cases flying apart, we also find two new, completely whole watches among the wreckage!

What's going on here? Where did these new bits of matter come from? Out of nothing? Perhaps so—but first we notice that these new particles came from the reduction in speed of the original two particles. In other words, the new particles were created from motion.

This is very interesting, because it doesn't support our everyday commonsense view of things. How substantial is matter—the book you're reading now, or the hand that holds it—if it can be created from something as apparently insubstantial as motion?

Astrophysicists tell us that motion is an expression of energy, and that the energy of the physical universe is of two kinds. There's positive energy, such as the energy that is locked up in matter. This is the energy we release when we set off nuclear bombs. It is also the kind of energy generated by the sun. But there is also a negative form of energy—we call it "gravity." It so happens that the amount of positive energy in the universe is equal to the amount of negative energy in the universe—that is, the total amount of energy in the universe adds up to zero. If we could gather all the mass-energy in the universe into one place, it would amount to zero too.[16]

Just how "substantial" is this stuff that is made from motion and energy, and that adds up to zero? Modern philosophy and mathematics have not been able to put away the inherent contradiction in the idea of motion discovered by the Eleatics, the ancient Greek philosophers who noted more than twenty centuries ago that a thing can move neither where it is nor where it is not.[17] Instead, they regarded Reality as without motion and unchanging—but this seems a bit extreme, considering that change is evident everywhere we look.

This argument has always reminded me of those who say that "all is one," even in the face of first-hand evidence that we live in a world of abundant multiplicity. As we shall see, our problem with motion is a psychological one. Anyone who has ever seen a movie can attest to the fact that "apparent motion" looks and feels like what we might otherwise call "real motion." Yet a movie is nothing more than a rapid series of still photos. "We're not **really** seeing moving pictures," we say. We'll discover, however, that no such distinction between motion and stillness—indeed, between oneness and multiplicity—can be made objectively. In fact, these are **not** two entirely separate phenomena. In later chapters, we'll explore how it is that such "opposites" must occur at once—and that it's **only** in this way that we can have an understanding of Reality that is free of contradictions.

The simple point I want to make here is that there are serious obstacles to overcome before we may attribute any substantiality to the physical world. Even G.E. Moore, the great champion of material realism, finally conceded in the end that he could not answer the Skeptics' doubts about the existence of materiality. Indeed, no one has satisfactorily answered the Skeptics to this day.

THE PHENOMENAL WORLD
REVEALS ONLY RELATIVE TRUTHS

Finally, in our effort to exhaust the *what* aspect of Reality, we will discover that by examining the external world, we can arrive **only** at relative truths—that nothing is Certain.

Let's consider yet another view of my cup. The cup sits upon my desk. But how can it **be** without a great deal of other stuff surrounding it—and, thus, defining it? At the very least, my cup needs to be surrounded by space. Furthermore, in order that we may experience this cup, we have to be situated away from it. If this were not the case with our objects, then we might not find anything ludicrous about an artist who sells plain white canvases that supposedly depict polar bears eating marshmallows in a snowstorm.

Our objects can **be** only in a dynamic relation with "other." Once we package up a small portion of the universe in concept—whether it be a physical or a purely mental object (e.g., an idea)—the only way we can actually have our object is in contrast to what it is not.

But "what-it-is-not" is necessarily an aspect of our object's actual identity—and, as we shall see, this aspect necessarily involves the rest of the universe.

In conceiving any object, then, we isolate and set it apart from what it is not. Therefore, any "truth" found in such a concept could be only relative and provisional at best. Like a seiche sloshing within a basin, or like an endless process of arranging and rearranging furniture within a room, relative truths replace themselves over and over, with (and to) no end. In other words, such truths will not satisfy the deep need of the heart. They are not Real Truth, and they do not provide us with Certitude.

The fact apparent to direct experience is that **any** theory (or concept), even a "theory of everything" (as scientists have dubbed some of their theories), necessarily leaves the *what* aspect of existence unresolved. What is the universe? What are atoms? What are sub-atomic particles? What is a person? What are life and death? What is reality? What is anything? As my Zen teacher used to put it, "Whatever you think, is delusion." Whatever conceptual answer we come up with is relative at best, and is never Absolute Truth.

MIND IS MOVING

We'll look at Mind and consciousness in Part II, but I would like to introduce a few points on the subject now, for these will help to clarify what's to follow.

The great mathematician John von Neumann concluded that, "from a strictly logical point of view, only the presence of consciousness can solve the measurement problem" and "the world is not objectively real but depends on the mind of the observer."[18]

Our problem of not being able to *see what's* going on occurs partly from holding to the commonsense belief in the primacy of matter over Mind, of an external world "out there" over perception. But if we insist on the primacy of matter over Mind, we will eventually be led to intractable problems.

For example, in his book, *Speakable and unspeakable in quantum mechanics,* J.S. Bell observed that:

> the most simple and natural…[way] in which quantum mechanics can be presented is called…"wave mechanics." What is it that "waves" in wave mechanics? In the case of water waves it is the surface of the water that waves. With sound waves the pressure of the air oscillates. Light also was held to be a wave motion in classical physics. We were already a little vague about what was waving in that case…and even about whether the question made sense. In the case of the waves of wave mechanics we have no idea of what is waving…and do not ask the question.[19]

It was physicist Louis de Broglie who first realized that not only were waves particles (bits of matter), but particles were also waves. As Nick Herbert wrote in *Quantum Reality:*

> New quantum facts destroy the once sharp distinction between matter and field. With two magic quantum phrases we can…[turn] matter into field and vice versa. It's beginning to look as if everything is made of one substance—call it "quantumstuff"—which combines particle and wave at once in a peculiar quantum style all its own.
>
> The world is one substance. As satisfying as this discovery may be to philosophers, it is profoundly distressing to physicists as long as they do not understand the nature of that substance. For if quantumstuff is all there is and you don't understand quantumstuff, your ignorance is complete.[20]

Distressing, yes. For starters, if everything is One, how do we explain the seemingly self-evident fact of multiplicity? What is this combination with a "peculiar style all its own," anyway?

There's a Zen story about two monks arguing over a flag that they see waving in the breeze. One monk said, "It's the flag that's moving!"

The other monk replied, "No, no. It's the wind that moves!" Wishing to get to the bottom of this question, they carried on in this way, back and forth.

When their teacher passed by and heard the monks quarreling, he said, "Mind is moving." What is this Mind the teacher referred to?

For those of us who would agree with the definition that the mind is what the brain does (a commonly accepted definition of "mind" today), consider how the brain is made of atoms, made of subatomic particles, made of…what? Motion? Energy? And what are motion and energy made of? What **is** the material world?

One of the central problems in quantum physics today is how it is possible for an arrangement of atoms to support consciousness (that is, how it can constitute a "measuring device"). But why the foregone conclusion that consciousness requires atoms? Does it make any sense to suppose that consciousness is constituted of atoms at all? According to scientists, the world remains in a state of superimposed possibilities until a measurement is made, thus determining which possibility is actual. **The act of taking a measurement collapses a potential into an actual.** And what is the act of taking a measurement? It's conception itself. As we shall see, measurement is an apparent alteration of Mind—an alteration that opens the door to uncertainty and probability.

What is known as "measurement" is a function of consciousness that collapses **perceived** Reality into **conceptual** reality, into mind objects.

We can devise a theory of everything and say, "*This* is Reality," "*This* is Truth." Or we can even say, "Mind is moving—that's the Truth, believe it." But our explanations don't cut It. It's only consciousness itself that cuts Reality—literally, as we shall *see*.

three

⤳(CONTRADICTION)⤳

> *The firmest of all principles is that it is impossible for the same thing to belong and not to belong to the same thing at the same time in the same respect.*
>
> —ARISTOTLE

> *A is not A, therefore it is A.*
>
> —DAININ KATAGIRI

A CONTRADICTORY WORLD

I once listened to a man who identified himself as a member of the Flat Earth Society. He noted that as a departing ship gets farther and farther away, it appears ever smaller and smaller, until we can no longer see it. He argued that this phenomenon was simply a matter of perspective, and nothing more.

The ancient Greeks knew better than this. They developed the idea of a spherical Earth, and pointed to the manner in which departing ships appear to descend beneath the surface of the Earth—first hull, then sail, then masthead—as evidence.

Believers in a flat Earth, if they would only examine the evidence, are faced with glaring contradictions. How are they going to explain this phenomenon of ships descending beneath the horizon? It seems paradoxical.

And there are other things just as difficult to explain. For example, when we travel south we find that the northern stars, swirling counterclockwise as they do, dip ever deeper under the northern horizon. Even the North Star will pass beneath the horizon once we cross the equator. Continuing south, after the sinking of the North Star, we'll find that all our familiar stars of the northern sky will depart, one by one, while the new stars of the south come to dominate the sky. But these stars swirl clockwise about a place due south that holds no star. Here we have another apparent paradox.

How to explain such phenomena? If you believe in a flat Earth, it's very difficult. If you truly believe the Earth is flat and you hang on to that belief, you've got numerous contradictions facing you once you take note of how ships actually appear over the horizon or how stars actually course through the sky. This and so much more will contradict your belief. But your understanding will deepen the very moment you admit that there are things in your experience that often contradict your beliefs. And once you accept the explanation that the Earth is spherical rather than flat, the apparent paradox unravels and your experience makes sense.

But until you do understand, the paradoxes, and the question of how to explain apparently paradoxical phenomena, remain. In your lack of understanding, you can only see a paradoxical world filled with things that "cannot be," yet which apparently are.

But does our world seem any less paradoxical than the Flat-Earther's? Is it possible to find situations where we might begin to notice some very odd and troubling things about the world? Where, like a Flat-Earther

that 2 equals 1. Any number times zero equals any other number times zero, for zero of anything is zero. Once we detect our logical error, the paradox is no more.

A nonmathematical example of a fallacy would be to conclude that a platypus is a bird since (1) they lay eggs and (2) all birds lay eggs. This is known as the fallacy of affirming the consequent. Again, the error is in our reasoning. Just because birds lay eggs, this doesn't mean that other egg-laying creatures—fish, reptiles, duckbilled-platypuses, etc.— are necessarily birds. We've simply made a mistake. There's nothing very mind-boggling about that.

Paradoxes of the "common sense is wrong" variety are not too disturbing either, once we see how we're mistaken in our background assumptions about Reality. But if we **do not** see the error in our commonsense assumption, the paradox will appear as "intrinsic and indelible" (Poundstone's words) as any genuine paradox.

The paradoxes faced by Flat-Earthers are of the "common sense is wrong" type. They differ from fallacies only in that the error involved is not necessarily one of logic, but simply one of not seeing what's actually going on. It's only when Flat-Earthers take note of how ships actually pass over the horizon that they suddenly run into apparent paradoxes. It's not merely a simple matter of logic, but an error in some tacit assumption they've made about the world. We can see the two conflicting premises that are involved. Each by itself does not pose any problem—but united, as they are in actual experience, they form a paradox. Stated in the form used above, the paradoxes will appear as soon as we attempt to hold the following two premises at once:

(A) The Earth is flat.
(B) Approaching ships appear from beneath the horizon.

Until we actually see what's wrong with our idea (in this example, of course, the error occurs in the first supposition), "common sense is wrong" paradoxes will appear as intrinsic and indelible as any of those of the "genuine" variety.

My Zen teacher used to ask me, "How can you live in a contradictory world? How can you walk down the sidewalk in a contradictory world? Nevertheless, you do." I didn't know what he meant at the time. But then I began to notice that, yes, the world was very strange indeed. Once it was brought to my attention, I began to see a paradoxical world appearing everywhere, just beneath the façade of my commonsense world.

"GENUINE" PARADOXES ARE "COMMON SENSE IS WRONG" PARADOXES

Let's take a look at some "genuine" paradoxes and see if we can find where we err in our commonsense thinking.

Though few of us boggle at watching ships appear and disappear on the horizon, most of us tend to boggle when we encounter the elastic quality of time. We used to believe—most people probably still do—that time ticks away mechanically, evenly. After all, we've invented machines—clocks—that steadily tick it off for us.

But we don't really understand time if we think of it in this way. If we closely observe our experience, we start to notice some strange things. Something appears quite paradoxical.

The first strange thing we notice is that the speed of light never varies. It's not relative to position or velocity. If someone approaches you at half the speed of light and shoots a light beam toward you, the light beam passes you at a certain speed—186,000 miles per second. And if someone else **receding from you** at half the speed of light shoots a light beam toward you, that light beam will **also** pass you at 186,000 mps.

When scientists first discovered this, it seemed to defy common sense. If you're driving down the freeway at 50 mph, and someone passes you doing 60, they'll appear to you to be pulling away at ten mph. People in the on-coming lane, on the other hand, if they're also traveling at 60 mph, will appear to close in on you at 110 mph. This is what we normally experience, and it's just what we'd expect.

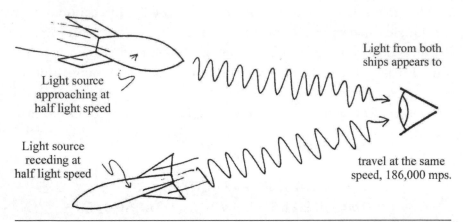

Light source approaching at half light speed

Light from both ships appears to

Light source receding at half light speed

travel at the same speed, 186,000 mps.

Figure 3–1

But our experience with light presents us with a picture like this: if, as you drive down the freeway, everyone else were traveling on light beams, they would all pass you at the same speed—i.e., the speed of light. Whether they are coming toward you or overtaking you, whether you greatly increase you own speed or come to a halt, or even turn around and go the other way, everyone would **always** pass you at the same speed. This is how light behaves. It seems very strange. How can this be? There seems to be a paradox here:

(A) Relative to me, the speed of an object is the sum or difference of my velocity and its velocity.

(B) The speed of light is constant, regardless of changes in my own velocity.

It literally took an Einstein to figure this one out.[3] Einstein was like the fellow on the shore who noticed how the ships crossed over the horizon—he noticed that certain observations conflicted with some of our commonsense ideas. Einstein wasn't the only one to notice the contradictions, but it was he who first realized that there was something wrong with our idea of time. If the speed of light is constant, he reasoned, then time (as well as space) cannot be constant, but must, rather, be dependent on the observer's motion.

This, too, seems counter to common sense. In fact, when Einstein proposed this new description of time and space, learned people proposed "thought experiments" such as the "twin paradox" in an attempt to show the absurdity of Einstein's idea. The twin paradox, a paradox of the "common sense is wrong" type, presents this strange situation: if time passes at different rates according to the motion of the observer, then if one of a set of twins travels at near the speed of light to Alpha Centauri, a little more than four light-years away, and returns to Earth, she will have aged only a few months—while her Earthbound twin will have aged ten years.

Today the "twin paradox" is no longer a paradox but an accepted fact, for it has been demonstrated that this is precisely what would happen in numerous experiments using jet planes and extremely accurate clocks. As Poundstone wrote, "The paradox lies in our mistaken assumptions about the way the world works rather than in the logic of the situation."[4] This paradox ceased to be a paradox the moment we began to hold a view that coincided with actual experience.

It's easy for us to still feel the twangs of this paradox even after we know what we're doing wrong. This is not unlike the fact that the tiers in the café-wall illusion (Figure 1–2) still appear tilted and tapered, even after we know the horizontal lines are parallel.

What I am talking about is understanding. As we begin to better understand what we're actually looking at—whether it's the flatness of the Earth or the constancy of time—we become aware of how our experience of things presents contradictions to us. These contradictions result from our having made very deep but quiet assumptions about Reality that, in fact, have no basis in Reality.

Once we begin to understand what we're really faced with, we might begin to see how many unfounded assumptions form the (incorrect) ground of what we believe.

For example, we understand very well that the Earth isn't flat, so we're not going to be sucked into a debate about what the edge of the Earth is like. But consider people of the early Middle Ages, many of whom believed the Earth was flat like a tabletop. Undoubtedly some of these people held heated discussions about what was at the Earth's edge. Some may have feared there'd have to be a huge waterfall somewhere. "Wouldn't the ocean drain off?" they might ask. "There must be some sort of lip there that holds the ocean back," others would reply. "Maybe there isn't any edge at all," still others might have said. "Maybe it's ocean forever? Or maybe after the ocean, some flat, barren plain stretches on and on."

To us, since we know that the Earth is in fact spherical, all such discussion now seems pointless and absurd.

Yet when it comes to certain ideas we commonly hold, particularly those that contain or engender paradox, we are just as much up in the air and wildly spinning as people of former Dark Ages. Substitute one of our contemporary issues and you'll see how easily we run off on an endless chain of wild speculation.

Instead of discussing what the edge of the Earth is like, how about discussing whether or not a two-month-old fetus is a human being? (It's best if you discuss this with someone with a slightly different angle on it than yourself.)

If, as I stated earlier, contradiction is an intrinsic part only of our **ideas** of Reality and not of Reality Itself, then it is precisely through our ideas that we fail to understand what is actually going on.

What I'm suggesting here is that whatever idea we have, we must examine it very carefully. And if we do, we'll find it **always** leads to paradox. Consequently, our ideas never quite explain what is actually going on. They do not (and, indeed, cannot) fit with anything we actually experience. They only elaborate and reify what we imagine and already believe.

If this is so, then we would do better to focus on what we actually **do** experience rather than on our **ideas** of what we experience. When we do, our problems with paradox and vain speculation simply do not arise.

THE DOUBLE SLIT

Let's look at some other paradoxes that regularly put us moderns into a spin.

The wave/particle duality in modern physics is common knowledge these days—I even saw a book of short stories titled *Light Can Be Both Wave and Particle.* But what does this mean, that light can be both wave and particle?

J. J. Thomson won a Nobel Prize in 1906 for proving that electrons are particles. Thirty-one years later his son, George, won a Nobel Prize for proving that electrons are waves. But waves are distinctly different phenomena than particles, for waves are spread-out things, while particles are point-like things. Yet George Thomson's discovery did not invalidate his father's discovery. They both stand firm today. As Nick Herbert put it:

> Whenever it's being observed, an electron always looks like a particle.... In between observations, the same electron spreads out like a wave over large regions of space. This alternation of identities is typical of all quantum entities and is the major cause of the reality crisis in physics.[5]

As Herbert says, it's not just electrons that behave this way. With the help of Louis de Broglie and others, we now know that **all** particles and **all** waves are a mixture of wave and particle. Says Herbert, "The world is made entirely of quons that behave like [an] electron."[6] ("Quon" is Herbert's generic term for all quantum entities such as protons, electrons, photons, etc.)

The question is, of course, how can anything be both wave and particle? It's a paradox:

(A) Electrons are particles.
(B) Electrons are waves.

Either possibility by itself poses no difficulty. But the two options appear to be mutually exclusive. It's not possible for both to be true at once, yet they are. It's a paradox—or, at least, it seems to be.

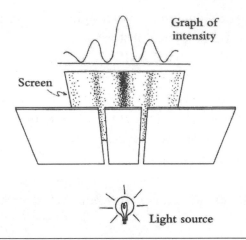

Figure 3–2. The wave interference pattern of light

Physicist and astronomer Sir Arthur Eddington observed that "no familiar conceptions can be woven around the electron." Rather, an electron is "something unknown, doing we don't know what."[7]

Consider the double slit experiment. Physicist Richard Feynman said, in his *Lectures* on quantum mechanics, that the double slit experiment is "a phenomenon which is impossible, **absolutely** impossible, to explain in classical ways [or, I might add, by common sense], and which has in it the heart of quantum mechanics. In reality, it contains the **only** mystery...the basic peculiarities of all quantum mechanics."[8]

This experiment, as Feynman implied, is the archetypal quantum mechanical experiment.[9] It's carried out by firing a monochromatic (meaning that all the photons—i.e., light particles—have the same energy) beam of light (or electrons, or any other sort of "quon") through a pair of narrow slits. Behind the slits is set up a screen to register the photons. Though they are particles when they arrive at the screen, the photons accumulate at the screen in a pattern indicating wave interference. This pattern appears as several alternating light and dark bands (see Figure 3–2). The bright bands (appearing as collections of black dots in Figures 3–2 and 3–3) on the screen are where the light waves coming through each slit are in phase with each other (that is, the peaks of some waves

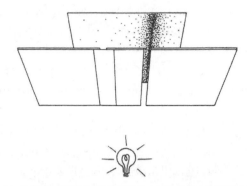

Figure 3–3. The diffraction pattern of light is compatible with a particle picture.

line up with the peaks of other waves, and the valleys of some waves line up with the valleys of other waves, thus reinforcing each other); the dark areas are where the light waves from each slit are out of phase (the peaks line up with the valleys and thus cancel each other out). This pattern is clearly indicative of light as waves.

If we block one of the slits, however, so that the photons can only reach the screen through one slit (see Figure 3–3), the interference pattern disappears and is replaced by a diffraction pattern. This pattern reveals that the photons are randomly deflected as they pass through the slit, striking that portion of the screen that is directly in line with the slit more often than the outlying areas of the screen. The likelihood of a photon striking any given area on the screen drops off in direct proportion to that area's distance from the centered line.

If the intensity of the light is relatively large—i.e., if the number of photons passing the slit is large—the illumination at the screen will appear uniform. If we cut back on the number of photons hitting the screen, however, we can see that the illumination is made up of separate points of light. This agrees with our understanding of light as particles.

So, when we open both slits we have an interference pattern. When we open only one slit we have a diffraction pattern. Common sense tells us that the wave interference pattern that occurs when both slits are open is the result of large numbers of photons streaming through both slits simultaneously. But that isn't the case. In fact, as mathematician and physicist Penrose explains, "each individual photon behaves like a wave entirely on its own."[10]

To test this, let's replace our screen with a photographic plate that will record each individual photon as it strikes the plate. Then let's set the entire contraption in a room of total darkness and shoot single photons—say, one per minute—through the two slits to the photographic plate behind. In the next room let's repeat the experiment, except that we'll shoot our individual photons through a single slit. Then let's go away for a week and let the experiment run.

When we come back and develop our plates, on the plate that recorded the photons passing through the single slit we'll find the usual diffraction pattern. On the one behind the double slit we'll find the interference pattern. But in our double slit experiment, we've only been shooting a single photon at a time—so what in the world could possibly be causing any interference?

To put it another way, if the photons in the double slit experiment were passing singly from the source to the photographic plate, and if they each arrive at the plate as a single photon, what was interfering with them that, over time, they gradually build up a wave interference pattern?

As it's commonly put, the individual photons somehow "know" when both slits are open. It would seem that the interference is caused by each photon interfering with itself when both slits are open. But how can a single entity pass through two slits? How can it interfere with itself? How can we have two without there being two?

Let me quote Penrose:

> The reader should pause to consider the import of this extraordinary fact. It is really not that light sometimes behaves as particles and sometimes as waves. It is that **each individual particle** behaves in a wavelike way entirely on its own; and **different alternative possibilities open to a particle can sometimes cancel one another out!**[11]

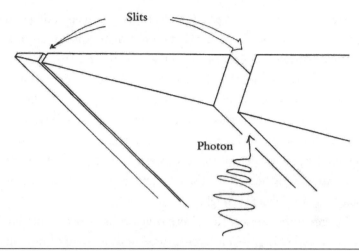

Figure 3–4. The slits from the photon's point of view. How can it make any difference to it whether the second slit, some 300 'photon-sizes' away, is open or closed?

To press the point home, we are asked to consider this strange affair from the point of view of the photon. Says Penrose,

> On the scale of the photon, if we take its wavelength as a measure of its "size," the second slit is some 300 "photon-sizes" away from the first (each slit being about a couple of wavelengths wide), so how can the photon "know" when it passes through one of the slits whether or not the other slit is actually open? In fact, in principle, there is no limit to the distance that the two slits can be from one another for this [interference] phenomenon to occur.[12]

Should we think of the photon as somehow splitting and going through both slits, but then reuniting before it hits the screen? Physicists developed an experiment to check this out—and came up with un-expected—and paradoxical—results. I'll quote Penrose again:

> As support for the view that the particle does not partly go through one slit and partly through the other, the modified situation may

be considered in which a **particle detector** is placed at one of the slits or the other. Since when it is observed, a photon—or any other particle—always appears as a single whole and not as some fraction of a whole, it must be the case that our detector detects either a whole photon or no photon at all. However, when the detector is present at one of the slits, so that an observer can **tell** which slit the photon went through, the wavy interference pattern at the screen disappears. In order for the interference to take place, there must apparently be a "lack of knowledge" as to which slit the particle "actually" went through.

To get interference, the alternatives must **both** contribute, sometimes "adding"—reinforcing one another…sometimes "subtracting"—so that alternatives can mysteriously "cancel" one another out. In fact, according to the rules of quantum mechanics, what is happening is even more mysterious than that! Alternatives can indeed add (brightest points on the screen); and alternatives can indeed subtract (dark points); but they must actually also be able to combine together in other strange combinations, such as:

$$\text{"alternative A" \quad plus} \quad i \times \text{"alternative B"}$$

where "i" is the "square root of minus one" (= $\sqrt{-1}$).… In fact **any complex number** [i.e., numbers containing the square root of negative one] can play such a role in "combining alternatives"!

…complex numbers are "absolutely fundamental to the structure of quantum mechanics." These numbers are not just mathematical niceties. They forced themselves on the attentions of physicists through persuasive and unexpected experimental facts.[13]

Part of our confusion in this matter stems from our lack of understanding about the role of consciousness. Detecting a photon (or **anything** else, microscopic or macroscopic), whether at the screen or at the slit, **is** to find it "there" and not somewhere else. But before we find it "there," it makes no sense to speak of it as having a location—for, if it doesn't appear in consciousness, it makes no sense to speak of "it" as existing at all. Yet our commonsense view of things is that they remain the same—and stay just as much "there" when no one is looking as when someone is.

To further examine the idea that things actually **do** abide "out there" when no one is looking, let's have a look at Schrödinger's strange cat.

A REAL LIVE DEAD CAT?

Physicist Stephen Hawking once said that whenever he hears of Schrödinger's cat he reaches for his gun. But how could a cat frustrate a man so? Schrödinger invented his cat (or, rather, his cat's situation) as a way to elevate to a macroscopic level the "two-not-two" phenomenon that lies at the heart of quantum mechanics. (It's not really necessary to rely on Schrödinger's cat to present us with so strange a situation, though. As we shall see, the sort of "strangeness" involved already occurs in our everyday lives all the time simply as the result of conceptualization—albeit on a more subtle level, and so we tend not to notice.)

Now, I will say at the outset that the cat in question is really a cat. I mean as we **commonly conceive** of a cat—that is, as we think of a cat, see a cat, hear, feel and smell a cat—that is, as a mind object. Like all cats, it's **always** either alive or dead. No one has ever experienced a living dead cat, and no one ever will. There is no cat, or anything else, that is **both** alive and dead simultaneously, because, in that we're dealing with a mind object, consciousness simply will not have it that way.

The problem we have with Schrödinger's cat is that we believe we're dealing with an actual, objective reality—i.e., a Real Cat, "out there"—and not with a mind object. But the fact is, we never deal with Real Objects, "out there," but **only** with mind objects. Hence, as we shall *see*, "live/dead" is not our problem. Our problem is "cat."

So, let's take a look at what Schrödinger and others find so vexing.

Picture a sealed chamber, so made as to not let any physical influence of any kind pass either into or out of its walls. In the chamber we've

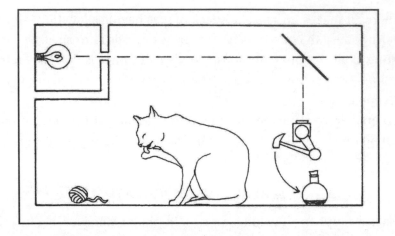

Figure 3–5. Schrödinger's cat—with a few Penrose modifications

placed a live cat, a vial of cyanide and a device that can be triggered by some random quantum event such as, say, the decay of a radioactive atom (this was Schrödinger's original version). When triggered, this device will smash the vial, thus killing the cat. In Penrose's version, the quantum event is the triggering of a photo-cell by a photon. The photon is fired inside the chamber to a half silvered mirror, where it faces a 50–50 possibility of either being reflected by the mirror to the triggering device, thus smashing the vial of cyanide and killing the cat, or being transmitted through the mirror and being absorbed by the wall behind, thus sparing the cat.

Okay. We've put a live cat in the chamber and sealed it shut. We have knowledge of the initial conditions of our quantum system, and the photon is sent on its way to the half silvered mirror—and the cat is now either alive or dead. Common sense tells us this is so. But, according to quantum mechanics, as in our double slit photon experiment above, until someone looks inside the chamber—i.e., until the outcome registers in consciousness and the cat is found either alive or dead—the cat is both alive and dead at once.

Why? Why doesn't it merely mean that the cat is in the chamber either alive or dead, as common sense would have it?

For starters, we could say it's because we're now dealing with a quantum system, which means we must use the mathematics of quantum theory—i.e., Schrödinger's wave equation. This dictates that, just as we needed the contribution of what seemed to be two mutually exclusive alternatives in the double slit experiment, we need both a live cat and a dead cat to account for the experimental facts. But, of course, this time we're not talking about a single photon going through two slits without dividing, but about a cat that is both alive and dead at once.

Earlier, Penrose invited us to pause and consider the import of the extraordinary fact that each individual photon behaves in a wavelike way on its own. **We are faced with the same situation here, except that now our quantum system has incorporated not a mere wave/particle, but a live/dead cat.** And just as the wave/particle evolves in a pattern of superimposed states until we make a measurement—i.e., detect it—thus collapsing this evolution into a definite outcome (a particle at *this* slit), so too does our poor cat seem to be continuing on in an eerie superimposed state of being alive and dead at the same time. And the superimposition of these states continues until we make a measurement—i.e., open the chamber and look—thus collapsing the two alternate possibilities into one reality, whereupon we find either a single live cat or a single dead one.

Something seems terribly turned around here. But what? There doesn't seem to be anything experientially wrong with the wave equation. The quantum theory, including Schrödinger's wave equation, is the most successful theory in all of science. It has performed flawlessly. As Penrose put it,

> probably most physicists would maintain that...there is now so much experimental evidence in favor of [the deterministic evolution of superimposed states, i.e., live/dead cats and wave/particles taking two paths at once]—and none against it—that we have no right whatever to abandon that type of evolution, even at the scale of the cat. If this is accepted, then we seem to be led to a very subjective view of physical reality.[14]

We've become like the Flat-Earther on the shore, noticing for the first time how ships actually appear over the horizon. We suddenly realize that we've held hidden assumptions about the nature of Reality. We've noticed something impossible about them.

We look for a way to explain what we think we see—but, in the case of quantum theory, our "explanation" leaves us more perplexed than ever.

If we do not accept the implications of quantum theory, such as live/dead cats, then we're forced to accept paradox as the basis of Reality, with all the confusion and uncertainty that ensues from such a rejection.

But if we do accept the implications of quantum theory, isn't a live/dead cat at least as much of a paradox?

Either way, we seem stumped.

On a deeper level, however, we will soon *see* that our problem stems from our lack of understanding of the nature of consciousness. Our difficulty with Schrödinger's cat arises from our commonsense view—i.e., that we take our objects of consciousness for absolutes. That is, we believe there's really something solid and clearly defined "out there" that persists through time. We believe in the solidity and persistence of live cats, dead cats, cups and even ideas.

But actually these things are just concepts, conceived things, thoughts. We can just as easily conjure up another kind of thought— say, a square circle. Strangely enough, though we can conjure up the **idea** of a square/circle or a live/dead cat, we cannot actually conceive what either term refers to. No one has ever witnessed a square/circle or a live/dead cat. No one ever will. Square/circles, wave/particles and live/dead cats simply cannot form as **objects of consciousness.** Go ahead, try to picture one of these "entities."

The point here is that discrete outcomes are **not** what we truly experience or perceive. Rather, discrete outcomes (e.g., particles, live cats, dead cats, etc.) are how we **package** Reality, how we **conceive** of things. These discrete outcomes, however, are not, in themselves, reflective of what's Really going on.

When we conceptualize a thing, we commonly overlook the fact that we've left something out—namely, the rest of the Universe. It's not quantum theory that forces paradox upon us. Rather, in holding to our commonsense view, our situation is like that of a Flat-Earther noticing for the first time how ships pass over the horizon. As long as we hold to an assumption that bare attention does not bear out, we remain confused. There's nothing wrong with what we *see;* our problem is with what we **think.**

THE THOMSON LAMP

Physicist Niels Bohr claimed, "anyone who is not shocked by quantum theory has not understood it."[15] Einstein apparently understood it; this was why he had many doubts about the implications of quantum theory and always felt the theory was incomplete. I suspect his reservations came from a feeling that the world **cannot** be paradoxical. Rather, it must be that we, in some fundamental way, habitually see it wrong.

But it's not just in the quantum world where things get sticky. Our experience of everyday things, if we attend to them carefully, will lead us to the same kind of paradoxical questions.

Let's take a look at the strange world of the Thomson lamp.

William Poundstone describes this lamp in *Labyrinths of Reason:*

> The "Thomson lamp" (after James F. Thomson) looks like any other lamp with a toggling on-off switch. Push the switch once and the lamp is on. Push it again to turn it off. Push it still another time to turn it on again. A supernatural being likes to play with the lamp as follows: It turns the lamp on for 1/2 minute, then switches it off for 1/4 minute, then switches it on again for 1/8 minute, off for 1/16 minute, and so on. This familiar infinite series (1/2 + 1/4 + 1/8 + ...) adds up to unity. So at the end of one minute, the being has pushed the switch an **infinite** number of times. Is the lamp on or off at the end of the minute?

Now, sure, everyone knows that the lamp is **physically** impossible. Mundane physics shouldn't hamper our imaginations, though. The description of the lamp's operation is as logically precise as it can be. It seems indisputable that we have all the necessary information to say if the lamp would be on or off. It seems equally indisputable that the lamp has to be either on or off.

But to answer the riddle of the Thomson lamp would be preposterous. It would be tantamount to saying whether the biggest whole number is even or odd![16]

Poundstone takes us into a logistical analysis of this dilemma, which ought to be enough to convince anybody that this paradox is indeed "genuine"—that is, he makes it quite clear that, since we'll not figure it out, this paradox is as "indelible" as any paradox could be. I will not pursue Poundstone's trail here, for, just as discussions of what the edge of the Earth might be like do not help us find our way out of the Flat-Earther's dilemma, so Poundstone's analysis of the problem only takes us deeper into the paradox. Rather, I want to see if we might be making some unfounded assumption in thinking that a lamp—not just the Thomson lamp, but **any** lamp—must be either on or off.

Poundstone's observation, that it must be on or off, seems reasonable enough. But if we look at this assumption more carefully, it will show how we're being misled.

Bare attention to actual experience shows us that, in fact, the "on" state is meaningless without the "off" state, and vice versa: each of the two states defines, and is defined by, the other. Each state is thus intrinsic to, and indelibly enmeshed in, its opposite. It is literally impossible for either state to exist without the other. Consider this question: Is the tree outside my window on or off? The question is meaningless unless we conceive of what an "on" tree and an "off" tree are. If we know that an "off" tree is a dead tree, then and only then can we determine whether the elm outside my window is on (alive) or off (dead).

This hearkens back to Schrödinger's cat. It seems absurd to think of a lamp as being both on and off, or neither on nor off, at the same time.

But what we miss is that being "on" always implies "off," and vice versa. When we say "it seems indisputable that the lamp has to be either on or off," we unwittingly imply the existence of an Absolute state of "On" or "Off." This paradox arises from our having posited these absolutes in our accounts of experience. If we would attend to bare perception, however, we'd notice that we do not experience absolutes anywhere—ever.

To put it another way, the idea that a lamp must be "just on" or "just off" without regard to its opposite state is an abstraction we've created in thought only. It's a concept, and therefore it's highly paradoxical. In practice—that is, in Reality—no such absolute abstraction appears.

I do not expect you to be convinced by these arguments. My point is simply that bare attention to actual experience shows us that the Real, Absolute, state of any lamp is in one sense both on **and** off—even though it's obvious that if we'd look at (i.e., conceive of) a lamp, it can't be both at once. Yet it **must** be both at once! Like a wave/particle, this is the fundamental "genuine" paradox we encounter with common sense.

Do we have any experience that might suggest that both the on and the off state appear together in some sort of superimposed state, which is neither on nor off?

Actually, we do. Whatever can be described as particle-like can also be described as wave-like, and all waves are subject to Fourier analysis, a powerful mathematical language used to describe wave motion. Called a "great mathematical poem" by Lord Rayleigh, the foremost British physicist of the day, Fourier analysis was invented by the French mathematician Joseph Fourier in 1823.

Fourier analysis allows us to express **any** wave—no matter how complex, no matter what its form, and no matter what its medium—as a sum of elemental sine waves. If a pure sine wave of sound, say, were being produced by some perfect musical instrument, it would be a steady tone unadulterated by tones of other frequencies. But musical instruments don't generally produce pure sine waves, but, rather, sums of sine waves—which accounts for why middle C on a piano sounds different from middle C played on a flute or a guitar. It also accounts for

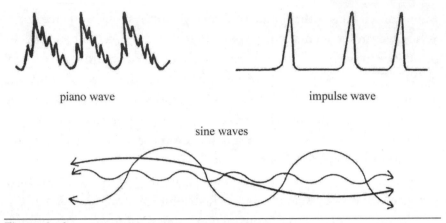

piano wave   impulse wave

sine waves

Figure 3–6

why human voices do not all sound alike. Different objects create different series of overtones. And it's these other tones that give each sound source its own peculiar signature.

Today modern electronic music can be produced by a synthesizer. A synthesizer can take, say, a piano's signature and break it down into pure sine waves through Fourier analysis. The synthesizer, producing the pure sine waves, then recombines them in just the right combinations to reproduce the piano's sound, more or less. (At least that's the idea. In practice, portions of the sine waves get left out, resulting in the sometimes unconvincing sound of a synthesized piano.)

The point that concerns us here is that these sine waves, out of which all waves are made, are infinite. As Nick Herbert puts it,

> An elementary sine wave is a smooth oscillation that goes on forever. Yet sums of these smooth endless waves are able to represent waves that are not smooth—waves with sharp corners, for instance—or waves that are not endless—short pulses, for instance.[17]

Like light waves that can combine to produce light and dark areas, sound waves can interfere with each other in ways that cancel out all sound. Thus, before and after the appearance of any sound or the light, interfering sine waves, which are infinitely long, must be canceling out completely. As Herbert notes,

In practice—music synthesis, for example—finite sine waves are necessarily used to synthesize instrumental wave forms, but strictly speaking each phase wave extends in its direction of motion across all space and vibrates eternally at one unchanging frequency....its technically infinite extent leads to a particular paradox.[18]

This is called Carvello's paradox. The paradox appears by this reasoning: a certain Dr. Carvello flashes a light in your eyes precisely at midnight. Before Carvello turns the light on, however, you put on a special pair of green sunglasses that, as you look in Carvello's direction, enable you to see the light even before Carvello turns it on. The idea is that since the light we see can be analyzed into a multitude of phase waves that are infinite in extent, it therefore follows that since these waves, being infinite, have always been with us, we should be able to detect them when they are filtered out from the light that we otherwise see under normal conditions. The reason we normally don't see a light before it's turned on, of course, is because the phase waves that go into making it up have always conspired—at least up until the point where the light manifests—to cancel each other out so as to reveal no light. But since these phase waves **are** "there," even though we don't ordinarily see them, it's just a matter of filtering out a number of them with the sunglasses so as to allow the remaining light through to our eyes. And therefore, with the special sunglasses, when we look in the direction of the source, even before the light is turned on, we ought to be able to see it.

That's the reasoning—but, of course, in actual experience, this never occurs. So why do we never see future events taking place as we look out upon the world with various light-filtering devices?

This paradox defied explanation for physicists until Hendrik Kramers and Ralph Kronig were able to demonstrate why such future-revealing sunglasses are not possible. It seems that if we made a pair of sunglasses that filter out all but a narrow band of frequency, they **would** indeed allow us to see into the future—except that, as Kramers and Kronig discovered, there is no material that can only **filter** light. It turns out that anything that filters light also disperses light, and it disperses it in just such a way so as to make the waves cancel out completely. The filtered light cancels out just as completely with the glasses as the full spectrum of sine waves cancels itself out when we do not wear the sunglasses.

Kramers and Kronig were able to debunk the paradox, but they did not debunk the Reality of infinite, superluminal phase waves. They are there; we just cannot see them directly. In other words, the "on" state and the "off" state of any lamp (or any object at all) **are** both parts of a single state, a single continuum that is broken only by our awareness (our consciousness) of "now they conspire to show on," and "now they conspire to show off."

And so the question remains: how is it possible to have two without there being two?

Just how this is not only possible but **normal** is where I want to focus our attention—for, whether we're speaking of a phenomenon that is wave/particle or an ontological state that is on/off, we always come up against this peculiar paradox.

This paradox of two that are not two runs through all that we commonly think we know or believe about the world, about Reality. Yet if we try to say whether the Real state of something is on or off, or wave or particle, then all of our pursuits down that road of commonsense questioning will be as vain as any speculations about the edge of the Earth.

In order to further penetrate this fundamental paradox of two that are not two, it might be easier if we first look at some "genuine" paradoxes of everyday life.

THE COMPLEXITY OF SIMPLICITY

Consider "simplicity." The word "simplicity" is defined as: (1) a simple state or quality...freedom from intricacy or complexity; (2) absence of luxury, elegance...; (3) freedom from affectation, subtlety...; (4) plainness or naturalness, as of behavior...; (5) lack of sense or reasoning ability, foolishness or dullness.

This is the *Webster Collegiate Dictionary*'s definition of simplicity. We approach most of our objects of consciousness just like this: we name them, then define their labels, and then take them in as if they were their labeled definitions. We first invent and then we reify. But nowhere in our definitions, as we can see in the definition of "simplicity" above, do we find any reference to that other aspect of simplicity. Just as "on" completely contrasts with "off," there is an aspect of simplicity that is completely contrary to what we think of as simplicity.

Of course, we all know what simplicity is, and no amount of clever manipulation of words can make simplicity out to be something other than what we know it to be. We know simplicity when we see it. But the thing we have to notice about simplicity is that if we pick it up, everything that **isn't** simplicity comes along with it. We must realize that, like either a wave or a particle, or on or off, we cannot simply have one aspect of Reality set apart, existing wholly on its own.

Let me illustrate this. I once spoke at a retreat in which people had gathered to examine, among other things, the idea of simplicity—more precisely, living the "simple life." One of the speakers was a woman who had spent a number of years living in the countryside in Wisconsin, raising a family. Many years ago, she and her husband decided that they would go off to the country and live a simple life. By choice they didn't have a phone. They didn't have electricity. They didn't have plumbing. They raised two children. And they surely led a simple life because, having so limited their activity, they made few demands upon their environment and upon the world.

Most of us at the retreat probably had a clear sense of what is meant by "a simple life." It meant living unpretentiously, humbly, and efficiently; above all, it meant being self-sufficient and not tied into or dependent on some massive, external infrastructure. Yes, we all knew what the simple life was, even those of us who didn't live so "simply." But as we discussed the idea of simplicity, we began to see a great deal of complexity in it. After all, here were our friends, clearly living a simple life—we all recognized that they did—but when it was time to do the dishes, they had to have already cut some firewood, which had to have already been cured and hauled into the house. They then had to stoke a fire, and pump their water from the well, and heat the water on the stove, and pour it into a pan, and regulate its temperature by mixing it with

though what I had learned gave me the background to write this book, what I really had to say to him could never amount to much in the end. I was able to reduce it to these two lines:

> Say "barrier" and two appears.
> *See* "barrier" and two cannot be found.

The longer we study our objects of consciousness, the stranger they become. Their boundaries become less clear as we gather more information about them. With more and more detail, the objects of discriminating consciousness reveal less and less of their own being.

For example, time, as the title of a book on my shelf states, is a "familiar stranger."[19] Though it is a common experience shared by nearly everyone, the more we focus on just what time is, the stranger it becomes.

If we say "barrier," or if we think "barrier"—and this is what we ordinarily think, what we ordinarily see—we see multiplicity. And the first thing we assume is: "I'm here." And immediately we must also assume that "everything else is out there."

If we say "barrier," if we think "barrier," if we see "barrier" in this ordinary way (which is not True *seeing* but, rather, mundane conceptualizing), then we see a world divided up. To see in this ordinary way is to see others as apart from "me." The world we commonly see ourselves living in is a world of barriers, a world of multiplicity.

But what is the barrier that divides us and yields all this multiplicity? If we are **Really** situated off from each other, how?

If we attend to the barriers—which must be there if there is to be multiplicity—and actually *see* what we've tacitly assumed were barriers, then we'd notice a very profound paradox indeed: we can't find two!

This has already been demonstrated with wave/particles, with ontological states such as on/off and live/dead, with concepts such as simplicity/complexity, and with ethical questions such as whether or not we are our brother's keeper. But how can this state of two-which-is-not-two be true of ordinary physical objects, such as stones and coffee cups and people?

Lao Tzu wrote in the *Tao Te Ching:*

> What is a vessel? If you take clay and shape it into a vessel, the function of the clay lies in the space that is absent of clay.[20]

Or, as a friend of mine once observed, "We all long to know the essence of things, yet we can only know what things are in their function."

If we look at Lao Tzu's image of the vessel and the clay, we might well ask, "What is the vessel?" Our commonsense view might have us say "clay," as if saying what it's made of accounts for it. But with such an explanation, clay becomes separated out and divided off from all that is not clay. This is not a perception of Reality but a concept, an **idea** of what is Real. And so, though we may believe "the vessel is clay," and may say so with full conviction (as though such an explanation were enough to account for things), it's still insufficient. For what is this vessel if it isn't also its function? If the vessel in question were, say, a cup, how could it be such if we exclude its function? And the function that is the cup is not merely in the clay but requires also the ambient space that is devoid of clay.

There seems to be a twoness here: a vessel, that is clay; and a vessel that is function. (Consider that, for example, the same function, which we still call "cup," also occurs in spaces that surround other objects made of plastic or paper.) And so we ask, "Well, what **is** this vessel, Really? What is the essence of this thing that we call 'cup'?"

We feel this question because we deeply want to know what things are in their essence—yet all we seem to be able to know is what things are in their component parts and their functions. In fact, apart from their functions, relationships and components, we do not seem to know what things are at all.

We are so used to thinking of things only in terms of their functions, components and relationships that we become all but blind to them.

We become so habituated to our concepts, and we assume those concepts so quickly and automatically and repeatedly, that we ignore the Reality that we've packaged as "cup."

We make ourselves coffee, and, somewhat blindly, while having a conversation, we reach into the cupboard and pull out a cup, almost ignoring it. We do this because, after all, "it's just a cup."

But **is** it, in fact, "just a cup"? Bare and careful attention to what the cup actually is yields quite a different answer. Thích Nhât Hanh speaks of it like this: if you really *see* what you're calling "cup," you must *see* the sun as well. For many eons the sun has supplied the Earth with energy, and it has helped to evaporate water into the atmosphere. The water has then condensed to form clouds, and, for many eons, rain has fallen on the Earth. You have to understand this if you truly *see* this cup, because over many eons of time, under the sun and with the rain, vegetation began to creep out upon the land, and mosses and lichens began to create soils, until eventually trees appeared. These trees get their nourishment from the air and the sun and the rain and the soil. And, being so nourished, the trees grew and produced wood.

And there was the person who thought to take some clay, and working with it a while, learned to shape it into many useful forms. And someone made a spinning potter's wheel and shaped the clay into "this cup." All of this thinking, all this ingenuity and activity, all of this is the "cup"—for we can't separate all *this* from what we call "the cup."

And someone fashioned an axe and took wood from the tree, and split the wood and dried it in the sun. And someone built a fire. Eventually someone thought of making an oven and baking the clay. All of this goes into our cup, too.

All of this is "cup" and must be included if we are to understand the Real Cup. The cup's identity is not just with itself; it is as much with everything that it isn't—with "other," with "not-self."

And where does this "other" end? **It doesn't,** because if we think about the sun, and the rain, and the many ways everything's connected, and if we think of the life that is dependent on the sun and on the rain, and the turning of the Earth, and all the stars (which are necessary, as we shall see, to define a "turning Earth"), and the whole cosmos, which is made up of stars and countless galaxies of stars—if we think of all this we can *see* that there is no end to "other," until the whole universe is taken in. The whole universe, everything and every thought, flows right into *this* cup. **That** is "cup."

So what we call a "cup" is, in a sense, the whole universe. It's everything, every thought, everywhere and everywhen. We can *see* this. We can experience this. We can *know* this directly.

But notice—once we've got everything taken in, like cosmologists who note that the total mass-energy of the universe sums to zero, we've got nothing at all. Nothing's being separated out. Nothing's being defined by consciousness.

Our commonsense, conceptual view of things is that they are separate from one another. Our mistake is that we take these relative, conceptual things for Absolute.

Let's consider your nose as an example. We commonly believe that a nose is a nose is a nose…and that's all there is to it. Its essence is forever **noseness**. Indeed, we might ask whether anyone can be so foolish as to believe otherwise.

But this is Absolute Nose. There's no such "Nose" anywhere in our experience of the world, existing separately, Absolutely and eternally. There is no such Absolute "Clay Vessel" either, or "You" or "Me" in actual experience. Bare attention to actual experience, prior to our conceptualizing, bears this out.

The nose you actually experience is a **relative** nose through and through. This nose requires, as an integral part of its very identity, everything it is not for us to experience it.

We simply cannot experience Absolute Nose, Absolute Cup, Absolute Anything. Yet Absolutes are what we commonly **believe** we find before us. We construct mind objects in the form of concepts, which we then unwittingly take to be Absolutely Real.

But we don't actually experience Absolute Things, because everything we experience is defined and implied by other things. A nose is defined by its location in space, the spaces and objects around it (eyes, cheek-bones, mouth, etc.), the light that illumines it, biology, geology, etc. Even events of history must be drawn into the framing of "this nose." And so it goes with a clay vessel, with "simplicity," with a photon, and with any object of consciousness conceived—including you.

THE INCOMPLETE LAW OF IDENTITY

So what does all this mean? It means that we commonly miss the hidden, dark aspects of our objects of consciousness. But they are always there. And they are there not as mere adjuncts **to** our objects but **as integral parts of our objects themselves.** They are intimate parts of our objects' very identity.

In other words, the law of identity is incomplete. In saying that a thing is what it is, we omit the obvious observation that a thing receives its identity as much from what it is not as it does from what it is.

As a result of missing the obvious, we make a lot of trouble for ourselves. When we unwittingly go only for the bright aspect of something, the dark, hidden aspect can sneak in and override (or at least undermine) the experience.

We tend to not notice, for example, that in pursuing security we make ourselves insecure. Or when we go for knowledge with a small k, we become more confused. Or when we look for substance, either in thoughts or things, we find nothing substantial. If we examine multiplicity we find Oneness.

But most of us don't go far enough to see the Oneness that backs multiplicity. Others do go far enough, but then get stuck in it, touting, "Oneness, Oneness, Oneness," denying the obvious multiplicity, the dark or reverse aspect of Oneness.

When we draw a concave line, we draw a convex one as well. And so, which is it? We say, "It depends." Of course, but what is it in Reality?

Let's take another look at the ambiguous figure we examined in Chapter 1 (see Figure 3–7). The outside of a soup can presents us with a convex shape. Do you see a convex shape in this figure? Or do you see a concave shape? Or do you actually see both a concave and a convex

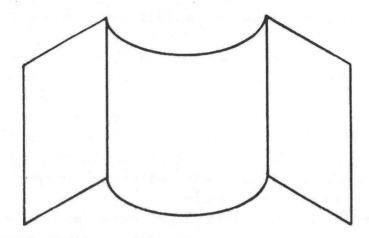

Figure 3–7. An ambiguous figure

shape? Is it possible that you see neither a concave nor a convex shape? After all, there's no actual curving surface here, but only this flat sheet of paper with certain markings on it.

We cannot say this figure is convex, for it's just as much concave; nor can we say it's concave for the same reason; nor is it both convex and concave—how can we have such a thing as that? Nor can we say our object is neither concave nor convex, for clearly concave and convex are there. What's going on here? What is this object, Really? Doesn't it have to be either concave or convex, or both, or neither?

This is a little strange, yet most of us are not likely to be emotionally jerked around by it. It's not enough to make us really feel the paradox— so let's step it up a little.

We're usually not perplexed while contemplating our baby picture, either. But if you were asked the question, "Is that you in that photo?" you might respond without difficulty, "Yes, that's me." Yet anyone can see that that's ridiculous, for you are a grown person and that's just a little baby in the picture. So should we conclude that it's **not** you? If it isn't you, then who is it? We might think, "Well, no, it's both me and it isn't me!" But how can that be? It's both you and not you? So then we may say, "Well then, it's neither me nor not me." But then, what are we talking about? Isn't there something, some object "out there," that got us started on this?

Nagarjuna, a brilliant Buddhist philosopher of the second century C. E., takes us through all these possibilities in many situations and shows us repeatedly that any object of consciousness—and this would include things like simplicity, electrons, lamps and coffee cups, as well as the subject in your baby picture—is never found to be (a) what it is, (b) what it is not, (c) both what it is and what it is not, or (d) neither what it is nor what it is not. All four of these possible ways of explaining experience do not stand up under scrutiny.

But then, what's going on? How do we explain experience, for obviously something's going on. What is it that forces this paradox, this "tetralemma," upon us?

If we can clearly *see* this paradox we can go through it. But first we must *see* that everything—every thought, every object, every idea that has some shape or form or definition that sets it off from other things—is like this. If we look at anything carefully, it reduces either to a paradox or to a tautology.

Such ontological problems are so serious that a Notre Dame conference convened in October 1987 to discuss the "Philosophical Lessons from Quantum Theory." A number of papers from that conference were collected and published in a volume titled *Philosophical Consequences of Quantum Theory* as an attempt

> to fashion an explanatory discourse with a view to producing an understandable view of our world. The ultimate goal is to construct a framework that is empirically adequate, that explains the outcomes of observations, and that finally produces in us a sense of understanding how the world can be the way it is.[21]

How **can** the world be the way it is? We need to understand our world. And our current lack of understanding is not limited to the quantum world. We need basic understanding desperately in all spheres of our lives, for when we look out upon the human world—whether politically, economically, morally, socially, spiritually, philosophically, ecologically or scientifically—confusion is most apparent. Clearly, nobody seems to *know* exactly *what* is going on. We need basic understanding to put our minds at ease, to know how to conduct our lives in a manner that fits, rather than defies, Reality. Yet we cannot comprehend how the world can be so strange.

With careful attention, however, we might find that it's really not quite so strange after all. Instead, we might find that we suffer all this confusion simply because we have not yet learned to pay attention to what we can *see* directly.

We've looked at our common object—a "cup"—the way Lao Tzu sees it. We've also seen the cup the way Thích Nhât Hanh sees it. We can even see the cup the way Schrödinger would have us see it (if we picture our cup as being in the chamber with the potential of being smashed like the vial). In the next chapter, we'll look at a common cup the way Nagarjuna sees it. And if we attend carefully, we'll go beyond Roger Penrose's observation that complex numbers are needed to account for quantum objects. We'll *see* that they are just as necessary to account for our most "ordinary" objects as well.

four

⫻(CERTITUDE)⫻

We used to think that if we knew one, we knew
two, because one and one are two. We are finding that we must learn a great deal more about
"and."

—SIR ARTHUR EDDINGTON

The greatest disorder of the mind is to let will
direct belief.

—LOUIS PASTEUR

The foolish reject what they see,
not what they think;
the wise reject what they think,
not what they see.

—HUANG PO

THE SCIENCE OF RELIGION

Science has always seemed to be about *knowing*, but in common practice merely sits upon belief.

Religion has always seemed, superficially, to be about believing. But any religion, for it to be in accordance with Reality, must actually sit upon Certain Knowledge.

Religion, contrary to what we commonly believe, and despite the myriad beliefs that have accumulated and become encrusted upon its name, cannot rest upon any belief whatsoever. Religion, if it is to be about Truth, cannot at base be about belief. It must instead offer deep, direct insight into the nature of Truth.

The word "religion" is rooted in the Latin *religare*, which means "to bind again." *Religio* is about binding with, or yoking together with, or reuniting with Truth or Reality.

Religion with a capital R thus invites and encourages us to *see*, to *know* Truth. Yet since belief, which inevitably grasps at mind objects, does not lead us to Truth, but to contradictions and paradox, then perhaps the correct function of Religion ought not to be supplying us with ever more beliefs. Instead, if Religion were to truly function according to its name—*religio*—it would have us **let go** of all our cherished beliefs, beyond any practicality they may temporarily or provisionally hold for us. Religion must be dedicated to the pursuit of Truth, Knowledge, and Certitude at literally all costs—particularly the cost of belief.

Religion cannot rest upon belief if it is to be True. And if it is not or cannot be True, why bother with it?

Regarding Religion, then, we're generally focused on the wrong stuff.

Joseph Campbell once noted that religion short circuits the religious experience by putting it into concepts. Thus, tragically, in the name of religion, we miss what can be *seen* directly, if only we'd learn to *look*. As my Zen teacher used to say, "Under the beautiful flag of religion we fight."

THE TETRALEMMA OF NAGARJUNA

In *The Emperor's New Mind*, Roger Penrose asks how we are to decide upon axioms or rules of procedure when trying to set up a formal system. He notes that our guide must always be our intuitive understanding of what is "self-evidently true."[1] But the question of what is self-evident is a touchy, difficult, and serious one. As we have seen, what is self-evident to common sense is often not what is self-evident to bare attention.

With bare attention we find we must include the *what* (or "what is it?") aspect of Reality, which common sense overlooks. We actually experience *what* as well as *this*, and so cannot legitimately reject *what*, or ignore it, or pretend it isn't present.

Most of us live out our lives with little awareness of the blatant contradictions that abound within our "knowledge" and beliefs (i.e., our concepts). Because we omit the *what* aspect, we make absolutes out of what would otherwise be understood as relative. Then we reify our concepts and sink ever deeper into contradictions.

We're so good at doing this that we rarely notice our mistake. We'd be better off just starting over and, with a concentrated effort, slowly and patiently scrutinizing our objects as they present themselves to us, rather than intellectualizing about how we've gotten it all wrong. This starting over is what we'll do in this chapter.

If we attend carefully, there seem to be only four possible ways to account for (i.e., conceptualize) objects of consciousness (i.e., experience). This is so whether our objects are exotic ones, such as photons, or pure mental objects such as the idea of simplicity, or common physical objects such as cups. These four ways form the four horns of Nagarjuna's tetralemma,[2] which, as we saw earlier, can be simply stated as follows: either (1) objects are themselves, or (2) they are not themselves, or (3) they are both themselves and not themselves, or (4) they are neither themselves nor not themselves. As we also saw, **none** of these four options explains or accounts for experience—yet experience remains.

We are now about to enter the world of direct experience. It's a world that common sense normally never allows us to venture into. It's a world of thoroughgoing relationship—and nothing but.

There's a crossover point from thinking of an object as Real and Absolute (i.e., that there really is an objective thing "out there") to actually *seeing* thoroughgoing interrelationship as Reality. Until we *see* thoroughgoing interrelationship as Reality, our mind stays locked in concepts and stuck in paradoxes. But when we *see* interrelationship as Reality, the mind is not so locked in concepts, and so doesn't get bogged down in confusion when called on to answer a simple question such as, "Is that you in your baby picture?" When Reality is not sought in mind objects that we imagine persist from moment to moment, but is *seen* in dynamic relativity, paradoxes no longer occur at all. The world no longer appears contradictory, and Certitude is realized.

SCRUTINIZING AN OBJECT

How **do** we account for the objects in our mind?

Let's take an object. This can be any entity that we "perceive" (conceive, actually) as existing apart from ourselves and other things—as a single, separate entity unto itself.

It could be any idea or thing. But to keep it simple, let's use a common everyday object that we unquestioningly take as being "out there." Let's say it's a plain porcelain coffee cup.

Of the four ways we might attempt to explain our experience of objects, the first one is our commonsense view of things. It's the straightforward assertion that all things, like this cup, are what they are—separate, real things situated "out there." They account for themselves, and they are self-defined.

Actually, this is more than just our commonsense view of things. It's also Aristotle's law of identity ("A is A"). The law of identity is one of the three "laws of thought," which were considered for centuries to be the foundation upon which all logical thought rested.[3] The law of identity, however, has gone into disuse in recent times—not because it's no longer widely held to be true, but because of its apparent indubitable truth. In modern times the law of identity has been seen as a tautology, so obviously true as to be utterly vacuous and useless. Today it sits on logic's uppermost shelf, nicely out of anyone's easy reach. Few people ever bother to take it out, dust it off, and examine it.

But when we do examine the law of identity carefully, as we did in the previous chapter (courtesy of Schrödinger's cat, Lao Tzu, and Thích Nhât Hanh), we find that it's incomplete. As a theory, it does not fully account for all that goes on in direct experience.

We will see exactly why shortly. Meanwhile, to help us better understand, it will be useful to express the law of identity in mathematical terms. (This will not involve anything beyond the number 1 and simple arithmetic.)

When an object appears in the mind, we conceive it as a solitary thing unto itself. Therefore we might express this mind object, this "thing unto itself," as the number 1. A mind object, no matter what it appears to be—a feeling, thought, or thing—always appears in the mind as a "one," a whole unto itself, separate from other feelings, thoughts, and things. It's a concept, a nugget, a unit—a singular thing of which we think and say, "I mean this and not that." It could be a cup, a banana squash, a galaxy, the idea of simplicity, or a vague pain in your knee. It could also be a photon particle, a drop of water, a bay, a lake, a live cat, a dead cat, or the idea of a live/dead cat. Each such object of consciousness is **always** experienced as a complete and singular entity.

Our commonsense idea, then—the law of identity—can be expressed as $1 \equiv 1$, meaning, "1 is identical to 1." In other words, our singular object, as it asserts itself in the mind, is what it is.

Thus, according to common sense, if our singular object were, say, a bunch of bananas, then when this object is asserted in the mind—as in, "Here are (+) bananas (1),"—the assertion of the object can be expressed as +1. If, on the other hand, we deny our object, as in "There are no (−) bananas (1)," the denial of our object can be expressed as −1. Negative one indicates a specific lack of a particular object—in this case, a bunch

Figure 4–1. Proposition 1: "A thing is what it is."

of bananas. Or, to put it another way, −1 affirms both the specific object (bananas, or 1) and the lack of that object (−, or −1).[4]

Intriguingly, however, denying a mind object automatically also asserts that object. Try to conjure up the negation of a mind object. It can't be done. When you think of the negation of an object—"no bananas"—the object, bananas, immediately comes to mind.

In dealing with mind objects, then, what −1 actually refers to is not a lack of the object of consciousness, but **what the object is not.** It refers to what remains of the Universe after our object has been taken into account. Negative one is the sum total of what our object is not, and it necessarily appears with, and in contrast to, our object.

In attempting to account for our mind object in mathematical terms, then, Proposition 1 (our commonsense view) gives us the first horn of Nagarjuna's tetralemma. It can be stated simply as "a thing is what it is," or:

A thing (1) is (+) what it (1) is (+);

or, a thing (1), in asserting itself (+1) in the mind,
appears to substantiate itself (+1);

or, a thing (1), as itself (1), is equivalent to itself (1);

or, $1 \times 1 = 1$.

Proposition 1, therefore, can be simply expressed as +1. The object (1) is the positive assertion of a thing as itself (+1)—our commonsense view.

As logical and as commonsensical as Proposition 1 may seem at first, Nagarjuna denies it as an explanation of experience—and he's right. Things that are merely themselves—i.e., things that are self-caused, self-defined, and self-identical—are not found in actual experience. In actual experience, no thing is ever manifested apart from "other"—that is, apart from what it is not, or without its background. As we saw in the previous chapter, **a thing is not merely what it is but also not what it is not.** All things receive their identity as much from what they are not as from what they are.

If things did receive their identity in the manner we commonly assume (i.e., merely from themselves, or merely according to their positive assertion), then all things would stand on their own with no need of any "other"; they would be permanent, unchanging, Absolute, and totally separate.

A corollary of Proposition 1 is that things persist through time; but close attention to actual experience reveals that there is no thing that persists in a single unchanging state, and that nothing is truly permanent.

On what grounds, then, can we justify denying what we directly perceive and upholding what we merely believe or conceive? Roger Penrose asks, "How can one be 'wrong' about what one actually perceives? Surely, one's actual perceptions are just the things of which one is directly aware, by **definition;** and so one cannot be 'wrong' about it."[5] He's quite right. We can't be wrong in our perceptions. We can, however, be wrong in our **conceptions,** in what we believe or imagine or reify.

The most profound of all commonsense mistakes is to take conception for perception. It's concept, not percept, that doesn't match Reality, and that opens the door for paradox and confusion.

Most people believe in Proposition 1, for it seems to most of us to be a simple and obvious description of direct experience. But, in fact, Proposition 1 is contradictory and irrational; it asks us to believe what we've made up (what we've conceived) and to ignore what stares us in the face (what we actually perceive).

Jeremy Campbell, in *The Improbable Machine*, says, "The mind leans over backward to transform a mad world into a sensible one, and the process is so natural and easy we hardly notice that it is taking place."[6] In a similar fashion, the mind leans over backward to transform an objectless world into a world of things and ideas. Through the functioning of consciousness, it turns a perceived world into a world of concepts. As Campbell notes, by "construing the outlandish as normal, we weaken our capacity to learn from experience."[7]

Proposition 1, then, is not a full report on experience. It's an incomplete description at best.

Since we must deny Proposition 1, that a thing is what it is, we might want to turn to some other idea, some other way of explaining how we experience things. A likely first move from our commonsense explanation is to simply negate it and state the opposite of what, until just a moment ago, had seemed most obvious. This brings us to Proposition 2, the second horn of Nagarjuna's tetralemma.

Proposition 2 states that "all things are what they are not." Put another way, our objects of consciousness do not account for themselves, but are other-caused and other-defined. In short, all things receive their identities from other things, from what they are not, from what is "not self." This view, the antithesis of the law of identity, is decidedly not a view of common sense.

Figure 4–2. Proposition 2: "A thing is what it is not."

This proposition can also be given mathematical expression in this way:

A thing (1) is (+) what it (1) is not (−);

or, a thing (1), in being defined by other (−1) in the mind,
appears only as the negative of itself (−1);

or, a thing (1), as the negation of itself (−1),
is equivalent to what it is not (−1);

or, $1 \times -1 = -1$.

This proposition suggests that, since +1 didn't work, −1 can account for the object of consciousness.

In other words, a thing (1), in being what it isn't, is −1. Proposition 2, therefore, can be simply expressed as −1, the negative assertion of the thing as itself. This is the antithesis of our commonsense view.

Not surprisingly, Nagarjuna denies this proposition as well. Things that are not themselves—things that are other-caused, other-defined and other-identical—receive their identity from what they are not. But we never actually experience such things. No thing is ever experienced as what it isn't. No cup is ever defined as "not a cup"; indeed, no object that is by definition not itself can be held by the mind. Consciousness simply does not allow it. We do not (and cannot) experience a cup as a mind object by virtue of it **not** being a cup. Proposition 2 clearly is not borne out by experience.

Since Propositions 1 and 2 each fail to account for our experience, and since each contains grounds for the other's denial, we may attempt to resort to yet a third way to explain our objects of consciousness. Proposition 3, forming the third horn of Nagarjuna's tetralemma, is arrived at by accepting **both** Proposition 1 and Proposition 2. It states that "all things are both what they are **and** what they are not." In Proposition 3, all things are seen to be both self- and other-caused, as well as

Figure 4–3. Proposition 3: "A thing is both what it is and what it is not."

self- and other-defined. In short, all things receive their identity both from self (i.e., "not-other") and other (i.e., "not-self").

Once again, we can put Proposition 3 into mathematical terms:

A thing (1) is (+) both what it (1) is (+) and (+) what it (1) is not (–);

or, Proposition 1 + Proposition 2

or, $(+1) + (-1) = 0$

Thus Proposition 3 states that $1 - 1$ can account for the object of consciousness.

But a thing (1), in being both what it is (+1) **and** what it is not (–1), amounts to zero. In other words, under these circumstances, no mind object is experienced at all.

As both the picture and the math above suggest, if we add these first two propositions together, the "thing" in question cannot be distinguished from its ground, nor can any identity of self or other be found. We lose our object of consciousness altogether.

With this "explanation," all experience would shut down, and all of space/time would be undifferentiated and unmoving. Since nothing could be discerned, it would mean that consciousness itself would have winked out.

Proposition 3 is thus clearly not found in direct experience. To say that things both are and are not is to say nothing at all. It's an "explanation" without meaning.

So what if we turn Proposition 3 inside out? This would bring us to the final horn of the tetralemma, Proposition 4. Here, in a feeble and last-ditch attempt to account for experience, we assert that "all things are neither what they are nor what they are not." This proposition argues that all things are without ground or cause—they are undefined. In short, Proposition 4 is that all things have no identity at all.

Again we can use a mathematical expression:

A thing (1) is neither (−) what it (1) is (+) nor (−) what it (1) is not (−);

or, the negation of Proposition 1 + the negation of Proposition 2

or, in effect, Proposition 2 + Proposition 1

or, $(−1) + (+1) = 0$

Thus Proposition 4 asserts that $−1 + 1$ accounts for our objects of consciousness. In other words, Proposition 4 can also be expressed as the number 0. No object at all can be accounted for under its terms.

Figure 4–4. Proposition 4 "A thing is neither what it is nor what it is not."

This proposition is also denied, for things that are neither themselves nor not themselves—things devoid of any identity or definition—are, once again, clearly not found in experience.

THE POINT OF DEPARTURE

None of these four propositions accounts for how we actually experience things. Nagarjuna, therefore, denies all four of these "explanations," despite the fact that, to our conceptualizing mind, there seem to be no other options.

Figure 4–5. Nagarjuna denies all four "explanations."

If we can't account for *what's* going on through these four propositions, then experience must not be accountable at all—or so it would seem.

Nevertheless, we **do** experience something—even though, as Huai-jang observed, to call it "something" doesn't hit the mark (see page 13).

Is there anywhere at all to go from here, other than frustration and despair?

Yes, there is. But it doesn't involve an explanation. It involves *just seeing what's* going on.

Let's look at our commonsense view (Proposition 1) once again. Here we do conceive an object. The object, according to this view, seems to imply itself as though it were Absolute. We can represent this mathematically as +1. But, as we have already noted, our sense of a thing is not merely that it is what it is, but also **that it is not what it is not.** This is the negation of Proposition 2, or the − of −1 (i.e., −1 × −1), which equals +1.[8]

So when we conceive an object—a thing or an idea—it doesn't merely sit there implying itself in the mind, as in $1 \times 1 = 1$. **It's also being implied by what it is not.** In other words, +1 is implied in the mind by an object's assertion of itself, +(+1), **and** by the denial of its negation, − (−1), which is again, +1.

We know that $1 \times 1 = 1$, or $1^2 = 1$ (pronounced "one squared equals one"). We also know that $(-1)(-1) = (-1)^2 = 1$. The square roots of 1 ($\sqrt{1}$) are thus both +1 and −1. This is just what we saw with our object. Its roots, and indeed its appearance, are found in both the positive and the negative assertion of itself.

The question that concerns us here, then, is what, in asserting itself in the mind, yields the negative root of our object (1)? In other words, what, when multiplied by itself, yields −1? What is the square root of −1 ($\sqrt{-1}$)?

The number mathematicians have assigned to this component of Reality is called i for "imaginary." But what **is** this i, this "imaginary" value that forces itself upon us in accounting for our experience of mind objects?

Mathematicians first came upon this number when factoring quadratic equations (i.e., equations with squares—for example, $a^2 = b$). They found that certain quadratic equations had no "real" roots; the only solutions involved the square roots of negative numbers.

At first, mathematicians called the square roots of negative numbers "imaginary" because they didn't seem to fit with any of the "real" numbers—or, for that matter, with any conceivable reality at all. When all the real numbers were strung out in a continuum running from negative infinity through zero and on to positive infinity, the number i would not fall on that line. It didn't seem to be a "real" number at all.

Though mathematicians could not directly conceive of what these "imaginary" numbers were, they were nevertheless able to work with them. By definition, they did have quite definite, conceivable properties of relationship. For example, by definition, $i \times i = -1$. It seemed, for a time at least, that imaginary and "complex numbers"[9] (i.e., numbers made of both real and imaginary components, which I'll simply express in the form $r + i$), though curious and entertaining, were no more than phantasmic inventions of the human mind. No one suspected, at first, that they had any relevance to the real world.

When mathematicians discovered that there was sometimes no way to arrive at real-number solutions to certain cubic equations without resorting to the use of complex numbers, they began to take imaginary numbers seriously—though they continued to think of these numbers as strange or inexplicable.

Eventually mathematicians, in particular Jean Robert Argand, showed that complex numbers could be arrayed as points in a plane, called the "Argand," or "complex plane," which is formed by the two axes of "real" and "imaginary" numbers set at 90° to each other and intersecting at zero (see Figure 5–2 on page 132).

After that, imaginary numbers were no longer thought of as imaginary (though the term has been retained for historical purposes). In fact, we now know that, far from being imaginary, imaginary numbers appear everywhere in the interrelationships that are the natural world of phenomena.

The importance of imaginary numbers lies in the fact that, since they are of a richer structure than real numbers, they can more easily solve certain kinds of mathematical problems. Indeed, physicists have used such numbers since the middle of the 19th century. In the quantum world, the world of Schrödinger's cat, such numbers are by far the simplest and most direct way to account for what's going on. This is why Roger Penrose made the point that "…complex numbers…are not just mathematical niceties."[10]

But it's not just in the quantum world where we may make use of these numbers. As we have seen, even our common experience of objects cannot be fully accounted for without the use of imaginary numbers.

Though imaginary numbers are no less real than "real" numbers, and though we can readily make use of them, there's still a sense in which these numbers might be called inconceivable, though not imaginary. They're inconceivable not so much in themselves (after all, we certainly do conceive them—if we couldn't, I wouldn't be able to write about them, nor would we be able to compute with them), but in that we cannot gather their **referents** into concept. In other words, they don't apply to mind objects. While we can conceive of one photon, one cup, or one idea, we cannot conceive of i photons, or i cups, or i ideas.

We might wonder why complex numbers ($r + i$, real + imaginary) seem to apply in some circumstances (e.g., they make computing the probable outcomes of evolving superimposed quantum states relatively simple), but not in others (e.g., they're not necessary in computing the probable outcomes of tumbling dice). This is because objects such as tumbling dice never leave (or incorporate anything beyond) the realm of conceived objects. The probabilities involved do not apply to anything that is not registering in consciousness. The same is true of light particles (photons), which appear as mind objects, and are accountable by real numbers alone.

On the other hand, accounting for light waves **does** necessitate use of complex numbers—and light waves **do not** appear as **objects** to the mind. Like live/dead cats, they're inconceivable. In other words, just as no one has ever seen a live/dead cat, no one has ever seen a light wave.[11] It's only when the imaginary component flees the scene (i.e., the i value goes to zero) that an "object" suddenly appears (i.e., is conceived) in the mind. At all other times the "object," unmanifested, remains lost in Totality.

There's nothing mysterious about any of this as long as we understand the nature of mind objects. **In terms of the way the mind works, there is no difference between classical and quantum reality.**[12] Physicists have long noted and puzzled over the fact that every conceivable thing has both a wave and a particle aspect to it. What they haven't understood is that every conceivable thing is a mind object—and **only** a mind object.

All objects of consciousness, then—though they ultimately cannot be accounted for by any of the four assertions of Nagarjuna's tetralemma—nevertheless reveal their true nature in a manner we can express as a complex number. All objects of consciousness can be viewed as being comprised of "real" and "imaginary" components—or, rather, relative (r) and inconceivable (i) components. In other words, all mind objects can be expressed as $r + i$, where r is relative (i.e., conceivable, conceptual, or *this*), and i is Absolute (i.e., inconceivable, perceptual, or *what*).

Thus can we mathematically account for both the nonconceptual, field-like **and** the conceptual, particle-like nature of things. We can account for both Oneness **and** multiplicity—the two-not-two aspects that, with close attention, our objects **always** reveal.

We are habitually taken in by our objects of consciousness—i.e., by r alone, our conceptual constructs. We assume that the endless parade of objects—contradictions and all—is Reality. Lacking an appreciation for inconceivability, we repeatedly look in the wrong place for Certitude, for a way to understand experience and the Universe. We look only to r, our conceptual understanding of things.

This has proven to be a great source of human suffering. Indeed, it may be our single greatest source of problems and pain.

part II

AT EASE WITH INCONCEIVABILITY

There's an ancient Chinese story of a king by the name of Chaos. King Chaos invited two guest kings, Brief and Sudden, to visit him. Brief and Sudden accepted his invitation, but, knowing nothing about Chaos, they had no idea what to expect. Upon meeting Chaos, however, their concerns were immediately put to rest, for Chaos proved to be a very efficient and elegant host. He was prepared to provide for their every need, and he cared for them very well.

The guest kings were treated with great dignity, respect and consideration. Indeed, they were so impressed by the fine hospitality of Chaos that they talked between themselves of how they might repay Chaos for his kindness. For a long while they could not decide what to do.

They had noticed, however, that there was something rather odd about King Chaos. He didn't have any holes in his head. There were no holes for his eyes, nor did he have any ear openings, or nostrils, or mouth. Chaos could not receive anything from the outside; he could not enjoy the world of the senses.

"Let us make him like us, that he might enjoy the world of the senses that we ourselves enjoy," said the kings. And so the two kings, Brief and Sudden, decided that they would bore holes into the head of King Chaos, so that he might enjoy the world of senses.

Not knowing the nature of Chaos, the guest kings heedlessly carried out their plan. Each day they drilled a hole, and Chaos did nothing to stop them. But at the end of the seventh day, after they finished drilling the seventh hole, King Chaos died.

five

⁓(C H A O S)⁓

When it is asked if the world is finite or infinite,
there is nothing in the mind corresponding to the
vocable world;...whatever we imagine is ipso
facto finite.

—THOMAS HOBBES

WHAT IS CHAOS?

We live as Brief and Sudden, and like them we don't readily understand Chaos. And so, like these two kings, we try to impose our sense of what is desirable and good—our sense of order—upon Chaos. We try to impose our sense of order upon the world "out there," upon what we take for Reality.

We do this so that we might make some sense of our experience. Yet all the while we do this, we overlook the fact that Chaos is doing just fine. Furthermore, we overlook the fact that Chaos is taking care of everything—beautifully, perfectly.

Yet, as it was with the guest kings who claimed to have never known such a wonderful host, this observation imparts little significance to us. Though Chaos provides for our every need, we do not take note of it. Rather, we try to shape what we do not comprehend into something of our own making.

So what is Chaos? If we contemplate Chaos, we may begin to notice that Chaos implies Oneness or Wholeness or Totality. This is very different from what we usually see.

To our ordinary, commonsense, fragmented mind—the mind that sees separate identities—the world appears as a sea of multiplicity. If we study it long enough, this world of multiplicity appears to rise out of Chaos, or total randomness. Yet this randomness nevertheless suggests Oneness. While we can see many different kinds of order in the universe, there appears to be only a single—and universal—kind of randomness.

Consider what happens when we randomize a simple system such as a deck of playing cards. If we take a fresh deck of cards, unwrap them, and lay them out one by one on the table, they will reveal a particular organization. In most decks, the cards will have been packaged ace through king in each suit.

If we shuffle the deck a few times and again lay the cards out, the order we once saw has started to break up. We might see a two, a three, a four, and then a queen. Some trace of the previous order clearly still remains, but it's becoming more difficult to distinguish.

At any time, however, we could reorganize the deck. We could lay out all the aces, then the deuces, then the threes; or arrange a red, red, black, black, red, red, black, black sequence; or place all the odd-numbered cards over here, and all the evens over there; or place all the face cards in one pile and all the numbered cards in another. We can arrange the cards into many, many patterns, and in each pattern nearly everyone could easily recognize order. It is precisely this potential for having many possible arrangements of order within a deck of playing cards that gives the cards their great gaming versatility.

But if we shuffle the cards, and continue to shuffle them until we can no longer discern any order at all, we then say that the cards are random. They are in what we might call a chaotic state—where each card in turn will have as likely a chance of turning up as any other card—and while they exist in such a state we will have no idea what card is going to come next. (Of course, if we do turn the cards up, one by one, we steadily gain information about what remains in the deck. To eliminate such a possibility, imagine turning up cards from an infinite deck.)

Take this randomized deck, one in which we can find no discernible order, and shuffle it some more. The result: continued randomness. Let's shuffle it still more. The result: an equal amount of randomness. No matter how much more shuffling we do, the cards will not become any more random.

There thus appears to be only one kind of randomness, although there are many different kinds of order. This is the simplest and most obvious manner in which randomness (and Chaos) implies Oneness.

Let's consider another example. Let's say we're going to cook up a large pot of spaghetti sauce. We sauté some chopped onions and minced garlic in a little olive oil. We add chopped green peppers and tomatoes, throw in some sliced mushrooms and spices, and put everything into a big cooking pot. Then we stir. As we stir, everything will become randomized into oneness. The spaghetti sauce will still be uncooked at this point, but everything will nevertheless be blended into a single whole.

But where is this whole, this oneness? If we take a very small sample from the pot, we might take out just a piece of onion or a slice of mushroom. If we just look at small samples such as these, we might very well conclude that there is indeed some order (i.e., organization) and not an undifferentiated oneness, for obviously the piece of onion doesn't resemble the slice of mushroom. But we're just looking at our sauce "locally," we might say. We're not looking at it as a whole. (This would be like picking up a totally shuffled deck and finding ace, deuce, three. We might conclude that the deck is ordered, but with such a small sample that conclusion must remain quite questionable.)

But any large sample of the sauce would look the same as any other sample—and each would look qualitatively the same as the whole. If we were to take a large ladle of sauce from a large pot, we're now likely to have a ladle that is not only representative of the pot as a whole, but also one that is essentially indistinguishable from any other ladle we might lift from the pot. In terms of content, each ladle of sauce would appear to be one and the same. Each would imply a oneness. Moreover, each would imply **the same** oneness, the same randomness, the same Chaos.

THE FRAGMENTED VIEW OF REALITY

We usually see everything from a particular point of view. We don't approach life from a perspective of Totality and Wholeness. We tend to come at life by seeing "myself over here" and "everything else over there." Our commonsense, fragmentary way of thinking—which is essentially a "matter before mind" kind of thinking—doesn't take Wholeness, or the Oneness that is implied by Chaos, into account. In so doing, it makes a number of assumptions that are highly questionable.

First of all, this view generally assumes that the human mind is contained in the brain—a most natural thing to assume if you believe matter is antecedent to consciousness. In holding to this belief, many people also assume that the mind is like a computer—that is, that the mind is packed full of a lot of information, and that it's "wired" for gathering and processing that information.

This is indeed a very common notion today. The human mind is often thought of as a machine that we carry around in our heads (some of us even believe that **we** are machines as well). Each brain is loaded with data, and has the ability to recognize faces and voices, to feel, to think, to calculate, and all the rest.

It's very natural for us to think like this when we maintain a fragmented, commonsense belief about Reality—that is, if we assume the world is primarily material, that it can be accounted for by real numbers alone, and that there's no subtle backing to the law of identity that pulls in Totality. But if we expand our view and see things from a broader perspective, we might begin to notice that the human mind is, first of all, **not** contained within the brain. Secondly, we might notice that little of the human mind has much to do with the brain (only enough to fool us—at least those of us who would rely solely on physical phenomena in our search for Truth).

Some simple experiments show us that conscious awareness is not merely "in the brain." Indeed, these experiments quickly reveal that conscious awareness is difficult (if not impossible) to locate at all.

Take your right index finger and touch it to your left palm. Massage the palm gently. You're immediately aware of both your finger and your palm. But where does this awareness take place? Is your awareness of your palm in your palm, in your index finger, or in your brain? Indeed, is it **anywhere** at all?

At first glance it may appear to be "out there" in your palm and finger and not in your brain at all—at least, this is what it **feels** like.

But suppose all the nerves in your left arm and hand were completely anesthetized. If you were to lay your arm on the table in front of you, and you then rubbed your left palm, you would not be able to feel your auto-massage. At least not from "within" the palm itself. But you can still feel your palm with your right index finger. This would seem to indicate that your conscious awareness is at least partly in your palm and partly in your finger. But it's also partly in your eye, because your eye also seems to register your left palm in consciousness. Indeed, any part of your body can be thought of as housing conscious awareness in one way or another.

It seems difficult to nail down just where conscious awareness ends and the body begins. Indeed, in some ways it's difficult to locate any difference between the two. Conscious awareness appears to be neither separate from nor identical with the body—nor does it appear to be both, or neither.

The problem in locating conscious awareness grows thornier when we go beyond the body to the world "out there." When you smell a flower, are you conscious of the smell in your nose, in your brain, or in the flower? Remove any one of the three, and no awareness of the flower's scent registers at all. Does this mean that your conscious awareness exists partly in the flower? Jointly in the flower, your nose, and your brain? It seems that with conscious awareness we become completely entangled in the world "out there," but in a manner much like "simplicity's" inseparability from "complexity." We seem to have multiplicity, yet we don't find multiplicity when we go looking for it.

If the scent of the flower that registers in conscious awareness does exist jointly with nose, brain, and flower, then where is that conscious awareness? Indeed, where is conscious awareness when viewing the light coming from a star a thousand light-years away? Is it with a star trillions of miles distant? In starlight that's nearly a thousand years older than you are? If your conscious awareness is entirely separate from that light and that star, then what **is** registering in conscious awareness?

It's very difficult to fix a boundary around anything, really. Where **are** the bounds of the body, let alone the mind? Is it really at the skin? Biologist Richard Dawkins has written extensively on how an organism's genes can exert influences that range beyond its body and into the environment—for example, a bee's genes can influence the colors of flowers.[1] When we begin looking where one thing stops and another thing begins, even in the physical world, we very quickly become entangled and confused.

Physicists, indeed, have a serious problem with consciousness these days. As physicist Nick Herbert put it: "Science's biggest mystery is the nature of consciousness. It is not that we possess bad or imperfect theories of human awareness; we simply have no such theories at all. About all we know about consciousness is that it has something to do with the head, rather than the foot."[2] Yet artificial intelligence researcher Danny Hillis claims "we're not even positive that thinking happens in the brain."[3] And if we're not even sure that thinking occurs in the brain, how much less is the evidence for conscious awareness occurring there—or in any locality at all?

As we shall see, when it comes to conscious awareness, we're very hard put to locate it anywhere—for, indeed, time and space themselves can be *seen* as directly resulting from the functioning of mental activity.[4]

The human mind has two aspects. One aspect we can duplicate very well in computers. It's what I'll call the calculating aspect (the *this* aspect). It's the aspect of mind that corresponds to the relative world—the world of thoughts and things. We can, in fact, make computers that far surpass the human mind at calculating. Experts in artificial intelligence note that while the human brain can make only 10 to 1000 connections per second, as of this writing some computers are able to operate at more than eight quadrillion (that's eight million billion) connections per second. By the time this book goes to print, the newest super-computer will be able to perform more than ten quadrillion connections per second.

But even though this is more than a trillion times faster than we can think and calculate, the same experts point out that they cannot get a computer to recognize the difference between a cat and a dog (something even a bird can do).

The second aspect of the human mind (or any mind)—the Awareness aspect—is Wholeness, Totality. Conscious awareness "has a component" that is unbounded and unborn. It's the *what*, the Absolute aspect of Reality.

Unlike a computer, the human mind is capable of perception, or recognition. This function (which is $r + i$, i.e., perceiving the two aspects of Reality at once) is the "unborn" aspect of mind. This is the very quality of mind that machines, being made things, cannot duplicate. It's this very quality of mind that makes us distinctly human—not as opposed to animals, but as opposed to machines, or things. It's this aspect of mind that can truly *see* Reality. Indeed, it's completely One with Reality.

We can distinguish three types of recognition that occur to the human mind. The most superficial is the mere naming of a thing (e.g., "banana squash"). This is mere labeling and categorizing. It's purely conceptual.

The second is also purely conceptual. It involves function and utility (i.e., subject and object), and it amounts to little more than being entangled within a bogus level of consciousness sometimes called "ego consciousness" or, in Sanskrit, *manas* (I'll say more about *manas* in

Chapter 6). This is a level of seeing where value judgments are made regarding the labeled objects of consciousness. At this level, what's "good" is what is seen as beneficial; therefore, a weed is merely a plant we have no use for. All such utilitarian concerns occur at this level of recognition. It's a level of grasping, craving, and wanting. We cannot become totally free of this way of seeing things as long as we maintain a view in which Reality is seen as fragmented. At this level we remain preoccupied with "the I creature" and what it owns.

The third level of recognition, however, is *just seeing.* Since it is pure perception, it's completely nonconceptual. In *seeing,* there's no object of consciousness as such. At this level, what is *seen,* or recognized, is Reality Itself, for our "object" has become interidentical with the Whole.

For the sake of clarity, let me expand on this point a bit. In his book *A Second Way of Knowing,* Edmund Blair Bolles lists recognition as one of the two basic forms of memory—the other being recall. Bolles notes that recall—e.g. who was the first president of the United States?—is conceptual. What we recall is always a symbol (or a series of symbols). On the other hand, recognition, says Bolles, does not produce a symbol. It's direct—e.g., you see someone at a party and, though you cannot recall their name, you recognize them.[5] But I am suggesting here that even this recognition involves a concept. Bolles' recognition is a form of level-two recognition.

Though these first two levels of recognition **follow** from perception, what I refer to as level-three recognition—i.e., pure perception—does not involve memory, or symbol, or representation at all. Pure perception is, in fact, without memory. It's True recognition, or *just seeing* without any conceptual overlay. It's immediate.

Thus conception involves the object alone (a condition, as we have seen, that cannot occur in Reality); whereas perception, as I use the term here, is not a mere fascination with an object, but the full vividness of *what's* going on—which necessarily involves the whole Universe.[6]

We can easily build machines that can duplicate the first function of recognition—and, possibly someday, even the second. But function number three cannot be built, for there is no way to set a boundary around it.

BOUNDLESSNESS

What does it mean to be boundless? In the story of King Chaos, Chaos could not take in anything from the outside. He had no sense openings, no "windows," to anything external to himself. He was thus complete, whole, unified, and boundless. In like manner, there can be nothing outside Absolute Oneness or Totality; It, too, is boundless.

Boundlessness, however, is impossible for us to grasp. It's mind-boggling if we approach it through our ordinary mind. Our ordinary mind sees only realities that are relative and therefore fragmented. Even in a physical sense, we have difficulty with boundlessness.

Modern science provides us with a very good physical example of boundlessness, however: the Universe. According to Einstein's general theory of relativity, the Universe is finite and boundless. We don't readily understand this—**finite** and boundless? If Einstein had said the Universe is infinite and boundless, it might be easier for us to nod and say, "I can imagine that."

To help us imagine a finite yet boundless space, scientists frequently use the balloon analogy. They ask us to step down one dimension and visualize our world as though it were in two dimensions rather than three. Then they ask us to picture our 2–D world as though it were curved like the surface of a balloon. And though it curves through a somewhat obscure third dimension (for the inhabitants of that 2–D universe), it demonstrates a surface that, like the surface of the Earth, is finite yet boundless (i.e., though finite, there's no edge to the Earth's surface).

Scientists would like to come up with some kind of image (though many of them have given up) that would frame what we know of physical reality into one phrase. We had such an image with the Newtonian paradigm; it likened the Universe to a big clock. Our clock-universe was a giant machine that, according to this image, was perhaps wound up at the beginning, but now just runs on and on. And from that beginning to the present day, and on into the indefinite future, everything within the Universe would ostensibly operate like this big machine. With such a view forming the backdrop of our collective psyche, we have even come to see ourselves as machines.

This model of the Universe (and much of what it implies) has fallen by the wayside now, though it will take a while for everyone to realize it. This image of the clock no longer suffices; it doesn't explain everything. It doesn't account for what we know of quantum theory's reality, or of the theory of relativity's reality.

Where will we find a concise phrase like "the universe is a giant clock" that will account for boundlessness? It's very difficult. Though the image of the balloon is helpful, it's inadequate because, as Sir Arthur Eddington pointed out, the more we learn about quantum physics, the more the Universe appears like a thought rather than a thing.

There are still other reasons why it's difficult to come up with a reasonably accurate model of Reality. Connected with the idea of boundlessness in relativity theory is the notion that the Universe is expanding. People often ask, "If the Universe is boundless already, how can it expand? And if it's expanding, what's it expanding into?" If we're still picturing the Universe as a balloon that is expanding, that image starts to come apart. For one thing, in the picture of the balloon, we're outside looking at it. In other words, we're borrowing an extra spatial dimension that we've not accounted for in our model. But of course this is not the case with Reality, for there's not (and, by definition there cannot be) anything outside the Universe into which it's expanding.

The expansion of the Universe, therefore, is not a matter of material expanding through space. What we're talking about is the idea that space itself (and time as well) are expanding. Space and time are both finite, and they both expand—or at least **appear** to expand.

But we can't get this image into a nice neat little phrase like "the universe is a giant clock." Instead, we tend to think that space is nothing—that it's empty. But actually space is not nothing—otherwise, how could it expand? There can be a boundary in space where "something"—space, at least—appears on both sides. But the physical universe in which we live doesn't have any boundary that divides it off from something else. There's no boundary because there **is** no "other side." Therefore, the *nothing* which supposedly lies "beyond" the Universe cannot be like the nothing we think of when we think of empty space. There's *nothing* outside the Universe. There's Really *nothing*, not even space. Not even time. Being Total, by definition, the Universe can have no boundary.

This is boundlessness. There's no edge, no border—*nothing* we can get a mental grip on.

BOUNDLESS MIND

Chaos implies boundlessness. In the story of King Chaos we see that, as far as Chaos was concerned, there wasn't any outside—there was *nothing* "out there" for which Chaos needed eyes or ears. Therefore, for much the same reasons that we need not concern ourselves with the perils at the "edge of the earth," he didn't need any windows to the "outside." In other words, there was no boundary that divided Chaos from anything. There was no "other" apart from him.

To be boundless means not to see something "over there" as if it were apart from you—or, indeed, as if there were some locality completely separate from "here." Unlike what we commonly believe we experience, there's no boundary that divides "here" from something else ("there").

We have called the relative aspect of Reality *this*. It's this aspect of Reality that we're consciously aware of experiencing. It's what we commonly think of as "real," and it leads us to believe we can account for our experiences of objects by way of real numbers alone (e.g., two warm-blooded ducks gliding into chilly water at 33.4 feet per second). It's what we have called the "r aspect" of Reality. We can also call it the "*something* aspect."

The Absolute aspect, on the other hand, is not commonly accounted for in our experience. It is, to the common mind, *nothing*, and it's often taken as being imaginary. Indeed, it can be represented by a number that mathematicians mistakenly took to be imaginary at first. It's this that we've called the *what* aspect of Reality, but it can be called the "*i* aspect" or the "*nothing* aspect" as well.

The notion of boundlessness implies Totality, which, in turn, implies Absolute—for if there's *nothing* "outside" (i.e., truly no outside), we're dealing with Absolute. And with Absolute there's *nothing* to compare It to. This accounts for why our relative, ordinary, commonsense mind habitually misses this aspect of Reality.

When seeing the world as a collection of parts, as we commonly do, we imagine boundaries dividing these parts. The mind that creates (or sees) these boundaries is ordinary mind.

The boundaries that we draw in our minds, such as making the distinction between "you" and "me," are very simple boundaries. We try to make our boundaries very plain and stark and bare because we want everything very clear and unambiguous. To the extent that we feel we've accomplished this, we feel we are intelligent or clever. We feel that we've accomplished something, that we've learned something, that we're getting somewhere.

But even though we may feel this, we ignore the quiet question that forever repeats at the very bottom of our psyche: "So what? What difference does it make?" Furthermore, we overlook the fact that we've taken the world down a notch. We've actually made it less interesting—and less understandable—though it might not appear so at first blush. Through our defining, our measuring, our dividing the world into "this" and "that," we make it more stark, more bare, and less in accord with Reality.

We don't readily notice this process, however, for we do it so quickly and by very small increments. Furthermore, all along the way we develop a deep fascination for our temporarily satisfying (which is to say, believed but ultimately dissatisfying) explanations of the world.

The boundaries we imagine to exist, given our fragmented view of Reality, are very simple straight-line designs. They are not at all like the boundaries we find in nature drawn out of Wholeness by Chaos.

Figure 5–1

When we set our boundaries, we generally imagine that they function in a manner much like those pictured in Figure 5–1. We want very definite lines around things, dividing things. We want to know gray as gray and black as black—and we want the distinction between them, where one stops and the other starts, to be very clearly shown. And because we want things very clear-cut, we commonly think that the boundaries between things and ideas must be like those in Figure 5–1—or, at least, that we can make them that way through our efforts.

This is how we imagine the boundary to be between things such as you and me, say, or between a road and a garden, or even between simplicity and complexity. As long as we do not scrutinize our boundaries carefully, it seems to be very clear where one thing stops and the other starts.

Furthermore, we see boundaries not just in physical terms, as between a road and a garden (where we might even conceive of a fence dividing them). We place equally firm boundaries in between our concepts. Our usual idea of good is of something quite distinct from evil; our idea of happy is entirely separate from our idea of sad. Even between our ideas of a road and a garden, our mind puts up a clear-cut, straight-line barrier.

But just what is that barrier? What is it that makes "road" separate and distinctly different from "garden"? Physically we might point to the fact that the road and the garden occupy different places; or perhaps they both existed "there" but at different times. Mentally we might refer to qualitative differences between what we call a road and what we call a garden. "A road is hard and sterile; a garden is lush and green."

But in fact there is (and can be) no Absolute Difference. As we shall see, there's nothing that divides "road" from "garden" except our thought. Philosophically we will be very hard put to pin down just what it is about time, or space, or qualitative states that make "road" and "garden" different.

BOUNDARIES OF INFINITE COMPLEXITY

Our main problem with *seeing* correctly stems from our desire to get the rules laid out and to devise visible and well-defined answers. We insist that the world must be "this way" or "that way." But we fail to *see* how in doing this, we make it all up—we repackage Reality and then believe in our package. And then we wonder why it doesn't make sense.

The world is not this way. Clear-cut lines, though necessary for our conceptual packaging, are simply not the way Reality is drawn. In Reality, for example, as we saw with Thích Nhât Hanh's cup, we do not simply end at our skins. Nor can we define our thoughts and things as though they're bounded within territories peculiar to themselves, like the areas of gray and black and white shown earlier.

We commonly imagine "the gray thing" to be divided from "the black thing" in a very simple, very straightforward manner—at least in regard to their grayness or blackness. We imagine things to exist within their own skins because we can't comprehend how Real boundaries are actually drawn. We can't deal with the fact that, in nature—that is, in Totality—boundaries are infinitely complex.

The fields of color in Figure 5–1 are in direct contact with each other. They butt up against each other; they touch. They are not intermixed or blended, nor do we have a sense that they are filled with each other, or that they **necessarily** imply each other, let alone the Whole.

This is very much not the case with things in Reality. In the Real World, all entities are forever totally intermixed—indeed, interidentical—with everything that they are not—and with Totality.

Like Thích Nhât Hanh's cup, things are intricately and inextricably merged with each other in an exchange of identity. And not merely with each other, but also with the Whole. In other words, in Reality black and gray (or this and that) don't exactly touch. The entire Universe settles into the interplay of exchanging identities, which is constantly occurring between them.

This sounds complicated and heady, but it's really not. A moment of reflection (or a moment spent looking at a color chart or the fading twilight) will reveal that there's no clear point at which one color "becomes" another. If you were to gradually darken a shade of gray, eventually it would appear to turn to black—but there's no one clear, objective point at which gray suddenly disappears and black appears. Things such as clear, objective points appear only in our **concepts**. (Indeed, this is precisely what conceptualizing is.) They don't show up in Reality.

Of course, on the quantum level, we **do** seem to find evidence of discontinuous changes—"that damn quantum leaping,"[7] as Schrödinger

Time Out!

In the next few pages, I will illustrate mathematically how it is possible that conceptual objects can appear to the mind while having infinitely complex boundaries (i.e., without being able to ultimately define "inside" or "outside" or "in between"). Please do not be put off by the math. Feel free to simply skip over any passages that seem too technical. You'll still be able to follow the essence of what I'm pointing out. If necessary, do the same in the few additional spots in this book where I use math. The math is meant to help readers who are mathematically inclined, not to deter those who are not.

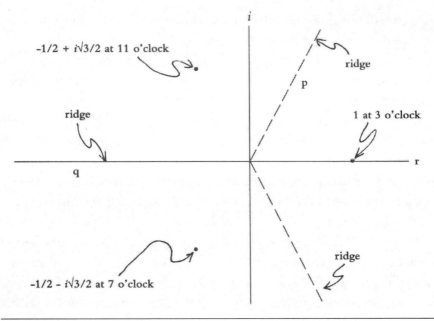

Figure 5–2

called it. But such discontinuity is the direct result of the functioning of consciousness, which quantizes the world into conceptual packets. If such discontinuity did not occur, consciousness would necessarily wink out, and purely *nothing* (i.e., the *i* aspect alone) would be happening—which is obviously not borne out by direct experience. On the other hand, if the world were **only** discontinuous, or accountable by the *r* aspect alone, then we would necessarily have endless paradox—which is precisely what we do appear to have when we hold to our common-sense views of the World.

This interplay of identities occurs not only with objects and colors and ideas, but even with **numbers**—precisely those things we think of as most discrete and clearly bounded—and with complex numbers, in particular.

For example, a degree-three polynomial, such as $x^3 - 1 = 0$, will have three solutions. X can have three values: 1, $-\frac{1}{2} - i\sqrt{3}/2$, and $-\frac{1}{2} + i\sqrt{3}/2$. Note the inconceivable term "*i*" in two of the solutions.[8] If we plot our three solutions for x on an Argand plane (see page 111 for a description of the Argand plane), these three solutions will appear at 3, 7 and 11 o'-clock (see Figure 5–2). If we look at the diagram of such a plane above, we might picture it as having three regions, each holding one of the solutions to $x^3 - 1 = 0$.

Some readers may be familiar with "iteration methods," which are techniques for finding successive approximations to the solutions of various polynomial equations where each iteration obtains a greater degree of accuracy than the preceding one. The ancient Greeks used a simple version of this technique for finding square roots. To begin, you make a guess. For instance, to determine the square root of 245, you might try 15. After multiplying 15 × 15, you arrive at 225. Then you try 16, and arrive at 256. Next you try 15.6 and get 243.36. Then you try 15.7, then 15.65, and so on. With each approximation, you slowly home in on the correct answer.

This simple version works for finding solutions to two-degree polynomial equations and square roots, but more powerful iteration techniques are used to solve higher degree polynomial equations, such as $x^3 - 1 = 0$. All iteration methods, however, use a homing technique.

Now let's look back at our diagram of the solutions to the equation $x^3 - 1 = 0$ (Figure 5–2). Note that this diagram will fit right over the figure of black, white, and gray (Figure 5–1) on page 129—but now the borders defining black, white, and gray appear as dashed lines. Think of the borders (the dashed lines) that define these three regions as ridges, and the solutions (the points at 3, 7, and 11 o'clock) as the lowest points in the three valleys formed by these ridges.

Now let's presume that we don't know any of the solutions to $x^3 - 1 = 0$ (as represented by the three plotted points), and that we are using an iteration method of successive approximations to try to home in on the answers.

We might assume, as we saw in our square root example, that if we make our initial guess at p (see Figure 5–2), we would move steadily away from the ridge and toward the solution at the lowest point in that valley—at 3 o'clock on our diagram where $x = 1$—with successive iterations. We might be astonished to discover, however, that if we begin

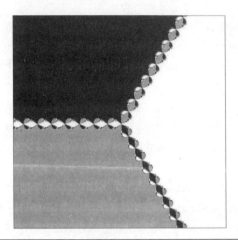

Figure 5–3

at p, we go to the solution at 7 o'clock, and that if we begin at q, we will home in on 3 o'clock.

Mathematicians discovered that something quite unexpected occurs at boundaries such as these. They are not the clean ridges we might suppose them to be, as with our pie pieces of black, gray and white in Figure 5–1. In fact, mathematicians have found that the boundaries between the territories of "black," "white" and "gray" are "ridges" of infinite complexity, each holding literally countless ridges and valleys of its own. Rather than clean-cut pie pieces, we find the following:

These boundaries are infinitely complex. And, far from the stark, bare, straight-line interfaces we conceive boundaries to be, what actually bounds **all** objects of consciousness—**all** things, **all** concepts—are "boundaries" such as these. Close attention shows them to be nothing like boundaries, exactly. If only we would scrutinize our objects (that is, all our things and thoughts) carefully, we would *see* this.

For example, let's consider a common object: a lake. Where is its boundary? What defines it?

If we don't scrutinize our object, this question will need no answer, and we think there's no problem—we think the lines are clearly drawn and our concept matches Reality. But where does the lake actually begin and end? How about this drop of rain now entering the lake? Is this part of the lake? What about the little stream that drains it, or the vapor that rises from its surface? (Indeed, under these circumstances, where **is** the surface?) And what about the water seeping down through the ground "beneath" the lake? Is that part of the lake? (If the lake were not there, the seeping water might not be there, either.) And is the water around this little pebble on the beach the lake? What about the fish and the microbes and flora of the lake?

What **is** the lake, exactly? What defines it? It seems a clear object at first, but as we go looking for it, it seems to have a very fuzzy definition.

If we don't ask, however, we can conceive "lake" quite clearly. Yet when we look for it, we find we cannot determine what defines it. Its boundaries are of infinite complexity. And this complexity continues on every scale.

For example, if we look closely at Figure 5–3, we can see that before the large field of black touches the large field of gray, everything else (which in this case is simply represented by white) jumps in between them. But now, before those little patches of white touch black, everything else in this little universe (in this case, gray) jumps back in between black and white. Gray refuses to be removed from Totality, we might say. But now, once again, before gray touches either color, the remaining color jumps back in between. No color can touch its adjacent color without whatever remains of the universe leaping back into the play. And so it goes. In this image we cannot have an entity, a part (such as gray) that does not imply and is not being implied by the Whole.

On any level, no patch of color is ever found to abut its neighbor, for whatever else remains of the universe is always found to appear in between. This pattern continues on down into the infinitely small with no end in sight. No color ever touches its neighbor, exactly, for at the last instant, just before we would conceptually define their boundary for them, they exchange what they are of themselves for whatever else remains of their universe.

The appearances of all entities, whether physical or mental, are just like this: even while they retain the distinct appearance of being things unto themselves, they exchange their identities with all else, and they

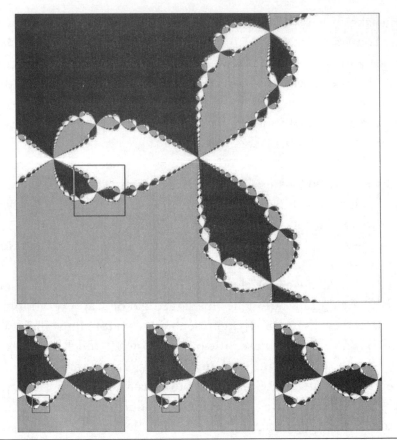

Figure 5–4. A series of images of Figure 5–3. Frame 1 is a blowup of the central hub in Figure 5–3. Frame 2 is a blowup of the box in Frame 1. Frame 3 is, in turn, a blowup of the box in Frame 2, and Frame 4 is a blowup of the box in Frame 3. In this series we can see that the interchange of identity at the boundaries is without end, for these trilobite shapes reappear at every scale, right on down into the infinitely small as we head toward zero.

become merged in intimate contact with what they are not. In short, they are interidentical.

There's no end to this exchange of identity at the boundary between what otherwise appear as distinct and separate things (or ideas). Such is the reality of all mind objects.

As all entities (whether things or thoughts) constantly exchange their identities with all that they are not, their boundaries appear to be infinitely complex. This is how all things appear in nature; this is how

Chaos (Reality, Totality) draws the lines. They are what we might call "no-boundary" lines. These lines are not at all like the boundaries we commonly draw (or imagine) between things.

We conceptualize things to be distinct in themselves and defined within definite, clear-cut boundaries. This is, as we say, "useful." But it's not Real. Not only that, it's not very interesting. Look again at Figure 5–1. Notice how simplistic it is—and how dead. Compare it with Figure 5–3, which is dynamic, and fascinating, and seemingly alive—even though it has been generated by things as seemingly "dead" as numerical values.

Upon careful scrutiny, all objects of consciousness exhibit the same sort of complexity as the images generated and displayed in Figures 5–3 and 5–4. Let's consider for a moment what James Gleick referred to in his book *Chaos* as the "grand-daddy" of all of these complex images, the Mandelbrot set. It has been called the most complex object in mathematics. (Actually, still more "complex objects" do exist in mathematics—though none have been studied quite so intensely.) Though the Mandelbrot set can be generated on a home computer screen from a modest list of instructions, Gleick observes that, "An eternity would not be enough time to see it all."[9]

The image, as it is drawn by Chaos, illustrates the paradox of two that are not two. Drawn upon the complex plane, it would seem that every point within the plane is either inside or outside the Mandelbrot set. Thus we have our two, and as we look upon the Mandelbrot set from a "God's eye view" (see Figure 5–5), we can clearly see that nevertheless there appears to be an "inside" and an "outside." But, as with any object comprised of both real and inconceivable aspects ($r + i$), the boundary between inside and outside (the defining line of discrimination) is literally infinitely complex. When we're away from the boundary, there's a clear inside and outside—yet as we move in for a closer look at our "boundary" (that is, as we gain more detailed information), the boundary cannot be found! It shows itself as being impossible to know, impossible to hold as an object of mind—even though its definition (the concept involved) clearly remains as an object to the mind.

Look at the series of images on the next few pages as we move in on the "boundary" of the Mandelbrot set.

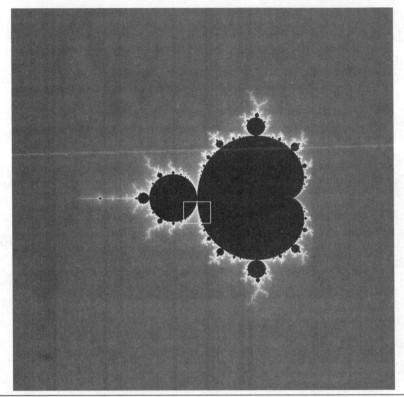

Figure 5–5. The Mandelbrot set is generated on the Argand plane by complex numbers iterating themselves over and over again.

Pick a complex number. Square it. Add its original value. Square that. Add its original value and square it again. For any given starting value, the numbers generated will either race off toward infinity or head for zero. If they go to infinity, plot the original value on the Argand plane in a shade of color that is dependent on how quickly this number leaves the vicinity of the "boundary." If it goes to zero, color it black. Black is "inside" the Mandelbrot set.

Most complex numbers quickly reveal whether they are "inside" or "outside" the Mandelbrot set. Those starting near the "boundary," however, will begin to turn and twist in their course before they take off for their ultimate destination. The closer to the "boundary," the longer they take to leave. Those numbers that are nearest the boundary will dance and weave for a duration approaching infinity.

In this frame we see the main body and shape of the Mandelbrot set in black, but as we move in on the defining "boundary," we discover its astounding complexity.

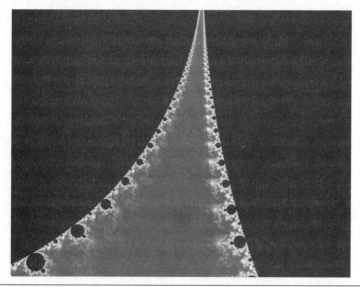

Figure 5–6. Here's a blowup of the box in Figure 5–5.

Figure 5–7. Here's a blowup of one of the nodules along the right-hand string of nodules in Figure 5–6.

Figure 5–8. A blowup of the large spiral in Figure 5–7

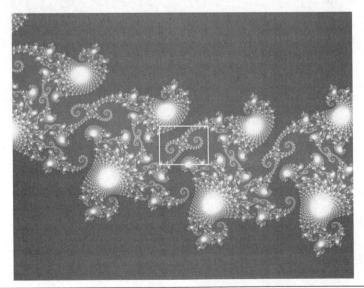

Figure 5–9. A blowup of the spiral arm in Figure 5–8

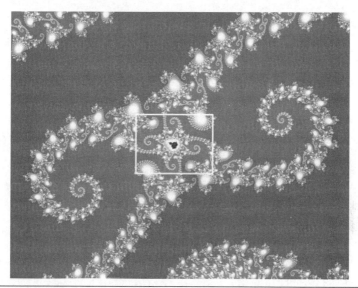

Figure 5–10. A blowup of the box in Figure 5–9 shows another, hidden, mini-version of the original shape.

Figure 5–11. Moving in one more time, again we find the endlessly repeating detail of that familiar shape.

INFINITE COMPLEXITY IN FINITE SPACE

The universe seems to be both finite and boundless. But the universe also contains a multiplicity of forms exhibiting, in some aspects, infinite complexity. But how is it possible to have infinite complexity occurring within a finite space?

This used to appear paradoxical to us. In fact, this question formed the basis for many of Zeno's paradoxes until we discovered, as in the series of ½ + ¼ + ⅛ and so on (which we encountered in our discussion of the Thomson infinity lamp) that, indeed, an infinite number of steps, or segments of time or space, can occur within a finite limit. The number series just mentioned will never exceed 1—even after an infinite number of steps have been taken.

Benoit Mandelbrot, the discoverer of the Mandelbrot set that we just looked at, coined a term for just this sort of complexity—he called it "fractal." Fractal, among other things, means "infinite line in finite space." But, as in the infinity lamp or Achilles gaining on the tortoise in each step by half of what remains after each previous step, how might we visualize such a thing as an infinite line drawn within a finite space? And where might we see such goings on in the world around us?

Let's consider the diagram below:

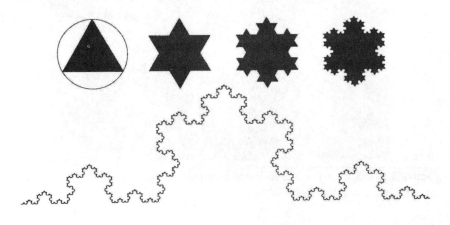

Figure 5–12. The Koch Curve

Start with a triangle having all sides equal to 1. Add new triangles to the middle third of each side and repeat at each scale. The length of the boundary of the emerging figure will increase at each scale by ⁴⁄₃. Thus, in the "final" figure, the length of the boundary is 3 × ⁴⁄₃ × ⁴⁄₃ × ⁴⁄₃ and so on, so it is of infinite length. Yet if we enclose the original triangle in a circle, we can see that neither the infinite line nor the area being enclosed by the infinite line will ever exceed that of the finite circle. In other words, we find the infinite merged and co-identical with the finite.

Another characteristic of Mandelbrot's fractal geometry revealed in the Koch curve is "self-similarity." With fractals, the same pattern is repeated again and again on ever-decreasing or ever-increasing scales—in other words, the pattern remains "self-similar" as we move from scale to scale. (You can see this in the trilobite pattern in Figure 5–4 as well as in the Mandelbrot set.)

But isn't this mere mathematical doodling? Where in our everyday lives, where in nature, where in Reality do we find examples of this sort of complexity, in which the finite commingles with the infinite, the simple with the complex, sameness with multiplicity?

We can see it in all the great intricacy of nature—in the veins and arteries of our bodies, for example. We see it in the manner in which the arteries leave the heart, dividing again and again until they are hair-like filaments running throughout the body. They suffuse the body in such a way that there is no cell that is more than a couple of cells distant from any capillary. Yet with these miles and miles of channels through the body, we find one system channeling blood in one direction, superimposed upon another system of vessels that channel the blood in the opposite direction. As scientists first learned of the complexity of the body, they were astounded as to how we could have such extreme complexity within the confines of a single living organism. Before the descriptions of fractal geometry, with its roots imbedded in $r + i$, there was no geometry that could account for such intricate complexity and multiplicity within the confines of a single, simple organism.

And of course, within the body we find not only the complexity of the circulatory system, but the complexity of the brain and nervous system;

and to that we must add the intestines, the kidneys, the liver, the lungs, and much more. And if we in turn choose any one of these systems—the lungs, for example—we find even more complexity superimposed upon all that I have already described. We find an enormous surface area inside the relatively small confines of the lungs—a surface area large enough to cover a tennis court. The inner surface of the lungs is placed within the body in such a way that all the necessary veins and arteries—some carrying oxygen-poor blood to each minute sack within the lung, and others quickly shunting the oxygenated blood away—are only micrometers from the lung's surface, the source of oxygen.

The branching out of the veins and nerves, and the various curves and lobes of structures of the body, are found to be exceedingly complex and yet self-similar at every level. Yet, of course, the human body is finite.

We find this sort of complexity not just within the human body, but in all life. We find it everywhere in nature. We can see it in snowflakes, with self-sameness of a different sort again appearing within infinite complexity—for while all snowflakes have their characteristic six- or three-sided patterns, we never find any two snowflakes alike. Though they are all different, we easily recognize them—for, in another sense, they are all the same. They all follow the same structure.

We see utter simplicity yielding infinite complexity in the way eyeless Chaos draws maple trees. All maple trees, all maple leaves, are alike, and yet each one is unique. Yet we can still recognize each leaf as distinctly a maple leaf. There's an aspect to each maple leaf and snowflake that remains from leaf to leaf and from snowflake to snowflake. It retains each leaf's "maple-leafness" and each snowflake's "snowflakeness." Like the trilobite shapes in Figure 5–4, they endlessly repeat themselves, yet no two are ever found to be alike.

And where else do we see such artless design? Everywhere. We see it in our fingerprints, in the spots on leopards, in the stripes on zebras, and in the grain of wood. Indeed, we can see it in our faces. These are all examples of "the same, yet different."

Infinite variety on a theme is what characterizes nature. And not just "living" nature. We see it in rocks, in sand dunes, in coastlines—indeed, we see it in all kinds of landforms that do not vary as we move from scale to scale.

Where we do **not** see such goings-on is in conceptual thought, in our commonsense straight-line ideas of how things are or ought to be.

We can see fractal geometry as the pattern of nature. We even see, deep within the fractal patterns, Chaos playing with that obscure duality of two and not-two, of simplicity and complexity, again and again. It's the two that cannot be as one, yet **are** together at one time and in one place.

But our commonsense view does not appreciate a world drawn with such complexity (or is it a complex world drawn with such simplicity?). Thus, through conceptualization, we make complexity not merely simple, but dead, and we become confused by what is not Real.

THE IMMEDIATE EXPERIENCE OF NO-BOUNDARY

We are prone to dismiss our most immediate experience. We do this in much the same way we forget about the air that fills the "empty" cookie jar. It's not that there's something mystical or vague about unmediated experience—rather, it's just that it's **so** commonplace, **so** immediate, that we overlook what we actually experience.

It was the immediate experience of conscious awareness that Descartes overlooked when he uttered his "cogito." He ignored the immediate experience of thought and instead made up—conceived of—"I think."

Consciousness precedes whatever it is we think we are referring to when we speak of self. To put it another way, consciousness precedes any experience of matter. Or, to put it in Descartes-like terms: *Consciousness, therefore "I" appear—along with "everything else."*

We never directly experience a time (or anything else) that precedes consciousness. Indeed, how could we? In fact, all our experience demonstrates that consciousness precedes matter. This is in direct contradiction to our commonsense view—yet with a little reflection we can directly *see* that this is so.

Every one of us experiences conscious awareness first and only then, possibly, matter—never the reverse. This is how we perceive Reality. Indeed, this is the only way we **can** perceive Reality—for how could awareness of matter appear without perception itself?

Despite this overwhelming experience, and not a single instance or piece of evidence to the contrary, our common sense tries to convince us that precisely the opposite is true—that matter precedes conscious experience and, indeed, that matter gives rise to consciousness. It's because we can easily conceive of (but never perceive) a time or place outside of our conscious awareness that we persist in holding this belief. Our fascination with our objects of consciousness, and our tendency to take them for Reality, only reinforce this belief.

But if matter were to actually exist prior to consciousness, then we'd be hard put to explain even the most common quantum event (as indeed we are, when we hold our commonsense beliefs). Why does the wave interference pattern change to a diffraction pattern as the result of an alteration in nothing except our state of conscious awareness? Why does observing the location of an electron make it impossible to determine its momentum? If matter actually preceded consciousness, these experimental results would be impossible to account for.

In the next chapter we'll explore more deeply just what consciousness is. For now, it's enough to note that it is a fact, demonstrated by direct experience, that consciousness is intimately involved in the creation of our realities. This is not analogy or metaphor; experimentation has shown empirically that it is so. Furthermore, if we simply attend carefully, we can *see* this for ourselves, directly.

There's nothing mysterious here. It's just a different focus from our commonsense mind. Consciousness has primacy over all phenomenal experience, including that of matter. Indeed, consciousness **is** the splitting of seamless Totality.

Yet this endless spewing out of things and ideas is only one aspect— the *something*, "*r*," or *this* aspect, we might say—of Reality. There's another aspect of Reality: Wholeness Itself. That is, the myriad things are not really many, for they are, in a sense, "One Thing." This other aspect—the *nothing*, "*i*," or *what* aspect—of Reality—is **uncompounded** Wholeness.

But, of course, there is in Reality **not** merely one thing, for there's also obvious—and immediate—multiplicity. So here's the paradox: there's Oneness, which is neither more nor less real than multiplicity. Thus, like "gray" and "black" in Figure 5–3, all entities merge and suggest the Whole, even while they remain in their "own" states.

The multitudes of separate entities, as they appear to our common sense, are real enough. But this, we should realize, **is** what discriminating consciousness is: the mental function that divides everything up. It determines, with physical objects and space, that "this is over here" and "that is over there," and with mental objects, that "that is this sort of a thing or idea," and "that is that sort of a thing or idea," or "this and that relate in these sorts of ways."

MERGING WITH YOUR OBJECT

When we examine any object of consciousness, whether it be mental or physical, the "rest of all that exists"—i.e., Totality, Wholeness—must enter into the picture. As long as we operate with discriminating consciousness and see ourselves **only** as a fragment—a part of Reality that is divided off and intrinsically separate from everything else—we can know only uncertainty, fear, misery, and that hollow, empty feeling of utter meaninglessness. Yet it need not be this way for us.

I cannot give you the direct *knowing* of that aspect which remains hidden from our ordinary conscious experience. I can, however, give an example that may remind you of this hidden aspect as it works in our everyday life. Let me tell you about my mother and lefse. (Lefse is a kind of Norwegian pancake made from potatoes, cream, flour, butter and sugar.)

Like all real boundaries, the boundary between my mother and lefse is infinitely complex. I witnessed this complexity as a child, though at the time I did not realize just what it was that I had witnessed. My eldest brother and his wife, newly married and inexperienced in the kitchen, had tried to make lefse on their own. Once they had put all the ingredients together, they discovered that they could not work with the dough. When they tried to roll it out it would stick to the board. When they tried to pick it up it would fall apart. They thought they had ruined it and were about to throw it out when, in desperation, they put in a distress call to Mom. I went along to see if I could be of any help. (I had a major interest in lefse in those days.)

My mother appeared on the scene like a midwife approaching a distraught husband. Rolling up her sleeves and taking sure command, she went to the huge lump of dough rising from the large mixing bowl in the center of the table. I can still see her as she put her hands upon that mound and said in a soft but certain tone, "Oh, it's just about right." Giving us a nod and a smile, it was clear that this baby would be spared. Quickly she dispatched her orders. It needed just a little more of this, and just another touch of that—and in seconds she was rolling out lefse and frying them up. Lefse appeared one after another, until soon the stacks were piling up under steaming cloths.

My mother's boundary was intimately connected with that of the lefse. The two merged, while nevertheless remaining separate. In fact, many things came together in that moment—not just my mother and the lefse. The dough had to be there, obviously. And though it was "just about right," my mother had to be there as well or there would have been no lefse. With my mother came the know-how—which, in turn, revealed that many other, previous and unseen events were also entangled in this happening of my mother making lefse. And within the dough were those who produced the ingredients, and those who trucked them to market. Within that dough were entangled the potato plant, and last year's harvest.

Yet while these countless hidden things came together in this event, it was nevertheless quite evident which was my mother and which was the lefse.

There's nothing mystical about what I'm trying to point to here. It's not a poetic metaphor or a Zen-like analogy. It's a simple, concrete example of that "other" aspect of Reality that must be accounted for if we would avoid irreconcilable contradiction. It's an example of someone actually merging in an exchange of identity with her object.

THE TWO TRUTHS

In Reality there appear two separate yet interacting things (subject and object, you and I, Mom and lefse, your eye and this page, good and evil, organism and environment, freedom and bondage). Yet we may be equally sure that, in Reality, there is also **no** such duality. There are two and yet there are not two. This occurs "at once" and in the same location. This fundamental duality is always occurring with conscious experience, everywhere and at all times.

With careful observation, we can see that the multiplicity of things always reveals unity. Conversely, if we begin with an assumption of unity, we can see that it will always reveal multiplicity. There is in Reality a merging of difference and unity.

This merging of difference and unity is the fundamental paradox that faces common sense (or, perhaps I should say, that common sense refuses to face). Nevertheless, we commonly witness countless examples of the merging of difference and unity every day.

We see a man tossing pizza dough, for example. At some point the man exactly understood pizza dough and became something that was not merely himself, but something interidentical with pizza dough. This is why he can swirl it in the air and stretch it out perfectly. Unless we know this merger for ourselves, we wonder, "How can a person do such a thing? If I tried to do that it would be a disaster!" If it seems amazing, it's only because we insist (without any conscious thought involved) that the man and the pizza dough are two distinct entities, that one must "use" the other, that they are two and only "relate" to each other, etc. The idea that there is a single entity appearing before us—that only one "being" is happening here—escapes common sense.

Human beings can do remarkable things, many of them far more remarkable than tossing pizza dough. We can acquire amazing skills and finesse. We can do acrobatics, perform a Prokofiev piano concerto, or devise inconceivable models of the quantum world that work flawlessly. But the reason we **can** do these amazing things is that we can merge with that object which common sense forever sees as "out there," separate and different.

If we analyze where the boundaries are between our objects and ourselves, we will eventually discover that we cannot definitely find them, either in time or in space. There is always this other aspect of Reality that is going on simultaneously, beyond the reach of common sense. Object and subject have merged; they're bound up with each other and with the Whole. At the same time, though, we can still see them as two, as separate.

JUST SEEING

Our problems stem simply from not *seeing*. We unwittingly and unnecessarily confuse ourselves—though we usually don't pursue our views far enough to see just how confused we actually are. Yet at any moment we have the power to *see* our situation for what it is.

Seeing is keeping our mouth shut—that is, keeping our discriminating consciousness, our ego, our intention, out of the question. *Just seeing* is all that is required to deal with our situation in a morally satisfying manner—that is, to deal with it in a way that is Total. It is our inability (or, perhaps more accurately, our unwillingness or refusal) to actually *see* that is our basic human problem.

When a question rises from the center of our collective, common-sense worldview, it often causes us to divide into separate camps. When this happens, it's not just us, but the world itself that exists in separate camps. We deal in concepts and we label things, and then we become attached to our views and our labels.

If we keep our minds quiet, however—if we still the inner dialogue that runs continuously within common human consciousness—we do not become confused.

If you're not convinced that we do indeed constantly chatter to ourselves, I invite you to try this simple experiment. Sit upright in a chair, feet flat on the floor, and do not lean against the back of the chair. Now, focus your attention on your breathing for five minutes. Just note your breath and nothing else. Go ahead—give it a try.

If you reflected upon your mental content during those five minutes, you noticed that your attention wandered from your breath many times. Your mind, in the span of five minutes jabbered and toyed with many mental objects.

This mental chasing about is not necessarily a bad thing. It is, in fact, normal conscious experience. In order to *just see* Truth and Reality, however, it's necessary to realize that the mind, when it's not under examination, jabbers to itself incessantly.

This is not to suggest that we should attempt to turn off the inner dialogue. Indeed, such a thing cannot be done—at least not directly. We cannot simply say to ourselves, "I'm going to stop that inner dialogue," and expect through the application of our will to succeed. This would be on a par with willing yourself to stop thinking about elephants.

So how **do** we stop the inner dialogue? We cannot stop it by expending energy and applying will or volition. Stopping the inner dialogue, making the mind quiet, only comes about through bare attention, from *just seeing*. The quiet mind, in fact, is none other than *just seeing* itself. *See* the situation. Whether it be the abortion issue, or the baby picture question, or a live/dead cat, set aside judgment and evaluation for the moment—either of the issue itself or of the views expressed by others— and *just see* the situation.

"Set aside" does not mean "ignore." We cannot ignore the abortion issue or our feelings regarding abortion, or the opinions of others regarding abortion—but, if we wish to *see* Truth, we should not get caught in an endless debate, either. This means that if our mind engages in a debate, we must *see* that our mind is engaging in a debate. If our mind grabs for an answer—"I'm siding with this view"—we must *just see* that we're siding with a view that is one of many. It's only through direct *seeing* that we may keep ourselves from going insane—that is, from losing sight of Reality.

Only when we *just see* can we *know* how the mind works. We'll begin to notice that to grasp at anything leads the mind into a quagmire of confusion. If we *just see* that the mind does this, we'll discover that it's in the *seeing* that we've already come back, that we've already merged with Reality (in the sense of "religio").

Awareness of Reality is *seeing*, not conceptualizing. The pivot point of volition is no more than the resolve to wake up, to *see* Truth, and to constantly return to *just seeing*. It's the suspension of judgment, and the disentanglement from the ceaseless, moronic chatter that characterizes our everyday minds. *Just seeing* is the end of the absurd belief that "*this* means that." It's the recognition that *this* can never mean "that," but is only the immediate *this*. It's simply to *see*, prior to concept.

six

ʍ(CONSCIOUSNESS)ʍ

[This awareness] is empty and immaculately pure, not being created by anything whatsoever.

It is authentic and unadulterated, without any duality of clarity and emptiness.

It is not permanent and yet it is not created by anything.

However, it is not a mere nothingness or something annihilated because it is lucid and present.

It does not exist as a single entity because it is present and clear in terms of being many.

[On the other hand] it is not created as a multiplicity of things because it is inseparable and of a single flavor.

—PADMASAMBHAVA

How is it that, apart from consciousness, there are no things in themselves?

Because the so-called "things in themselves," if examined in the light of reason, do not exist at all.

—HSÜAN TSANG

*Physics is the study of the structure of conscious-
ness. The stuff of the world is mindstuff.*
—SIR ARTHUR EDDINGTON

THE STRANGE FAMILIAR

What is consciousness? The word is frequently tossed about these days,
but what does it refer to?

As we've already seen from experiments (e.g., the double slit) and
thought-experiments (e.g., Schrödinger's cat), our acts of detecting and
measuring—seemingly primary acts of consciousness—do little to clarify *what's* going on. Indeed, the strange results of these experiments
seem to invite us into a dreamlike reality. The wave interference pattern
on the wall changes to a diffusion pattern because we've gained knowledge of which slit the photon went through. How is it possible that the
mere shift in our awareness can cause physical phenomena to change?
Indeed, what is consciousness, anyway?

Consciousness, it seems, doesn't merely stand outside Reality and
objectively perceive what's going on. Rather, it seems to interact with
everything on a very deep, basic level. In fact, as we've just noted, it
seems, somehow, to **create** realities. Consciousness seems essential to
and intertwined with objects and events, and all these objects and
events seem so intertwined with each other, that everything does indeed
seem to fit together as though it all were woven into a single Whole.

And yet multiplicity also exists. It's obvious that it does.

Physicists have uncovered some truly strange things regarding consciousness—things few people expected to find. But consciousness has
always been rather vague and mysterious. Our perennial confusion surrounding the term "consciousness" is well illustrated by Julian Jaynes in
his book, *The Origin of Consciousness in the Breakdown of the Bicameral Mind*. Jaynes mentions eight different ways we've historically at-
tempted to account for consciousness. In every instance, however, it
seems we began with the assumption that consciousness either origi-

nates in, or is associated with matter. It's either in matter originally; or in protoplasm—perhaps it resides in a collection of ganglia at the top of the spinal column; or it evolved with us as our physical form evolved; or, as we evolved, consciousness came forth suddenly; etc. Our attempts to explain consciousness have even gone so far as to include behaviorism, which, more or less, openly denies the existence of what Jaynes refers to as consciousness. As Jaynes points out, however, with every definition there arise intractable problems that prevent any adequate understanding of what consciousness is and how it originates.

According to Jaynes, the problem of consciousness remains to be solved, so he offers us another solution that departs from our old matter-oriented approach. Noting that "consciousness operates only on objectively observable things," Jaynes suggests that consciousness is the "invention of an analog world on the basis of language."[1] In other words, consciousness is the packaging of the world in the mind, but on the basis of language. But Jaynes' definition doesn't seem powerful enough to explain the apparent intermixing of consciousness with the world—or, indeed, the actual triggering of the world, which, according to our quantum experiments, seems to be what's happening with consciousness.

It seems that what Jaynes refers to as "consciousness" is what Buddhists have called *manas,* or ego consciousness. It's a bogus form of consciousness, however, for it presupposes the existence of the subject—"I."

Manas specifically denotes that form of consciousness which consists of our conceptual thought constructs. This is the consciousness of common sense, the consciousness of self and other. It's the consciousness we infer in a phrase such as "I'm aware of...." It's also the consciousness of contradictions, omissions, and limitations, for most of what is experienced escapes (or is rejected or ignored by) *manas.*

Manas appears to stand apart from the world; but, as we have seen, we cannot actually find any entity that **can** stand apart from the world. Furthermore, Jaynes, like most of us, presupposes that the objects of consciousness are really "out there." Hence, Jaynes' definition of consciousness cannot explain why things (and ideas) lose their substantiality as we approach them.

It's only when consciousness is seen as antecedent to matter that our problems with consciousness cease. And just as consciousness must be *seen* as antecedent to matter, so too, it must also be *seen* to precede language. Indeed, consciousness is the basis of language, not the other way around.

Unlike Jaynes, however, most people today (and, I suspect, the majority of scientists) still hold that consciousness is somewhere, somehow, within these few pounds of flesh. "The mind is what the brain does" is the current definition of consciousness. But as long as we continue to attempt to account for consciousness in this manner, we'll never realize its nature.

Our commonsense view, which assumes the origin and place of consciousness to be in matter, or else in language, simply cannot account for consciousness. These notions do not come close to resolving the Reality crisis brought on by quantum physics. This, no doubt, was why writer-physicist Nick Herbert, more or less speaking for scientists in general, claimed in 1985 that we do not have any explanation for consciousness at all.[2] This situation remains true today. It's because of their utter faith in the primacy of matter that physicists, at least the few who would have an ontology, are now baffled by consciousness. Consciousness has finally entered into their experiments, and they don't know how to deal with it. Given their tacit assumptions of a world external to Mind that is filled with separate and distinct entities, there's no way of accounting for consciousness—or, for that matter, the world.

But let's pause for a moment. See if you can notice that Mind—and by that I don't mean merely "your mind," but simply raw happening, *this*—neither comes nor goes. Our minds constantly flit about, of course, often dramatically and far more than we may realize. Our **objects** of consciousness come and go endlessly, too, never resting for a moment. But we don't actually experience Mind Itself "originating" anywhere, or anywhen. As far as direct experience is concerned, Mind—*this* Awareness—is ever-present and immediate. In other words, we never directly experience *nothing*. It always appears as though *something* is going on in actual experience. But what is it? It's like a strange familiar.

THE PRIMACY OF MIND

Though we can't confirm through conscious, or conceptual, awareness that an external, objectively real world exists (as we have seen, our conceptions can always be shown to be illusory), direct experience **does** confirm pure, objectless Awareness—i.e., perception, that *this* is going on. It's from this perspective that I argue for the primacy of Mind over matter.

When Mind, or objectless Awareness—pure perception—is seen as antecedent to rather than the consequence of matter, then it becomes unnecessary to explain consciousness in terms of evolution. Nor does it make sense to think of it as having originated somewhere, for now Mind can be *seen* as the originator, instead of the product, of place and time.

If we are to say, as common sense would have it, and as Jaynes does, that there's an "origin" of consciousness—as we would think of "my consciousness"—then we might also conclude that there must be a time that precedes consciousness, since we can easily imagine a time that did precede "our consciousness." But this is merely a concept. No such time is actually given to direct experience—i.e., no such time is available to perception. Each of us lacks any direct experience of a time—past, present, or future—outside of immediate conscious awareness. No one is ever conscious of not being (or not having been) conscious. Indeed, such awareness is clearly impossible; to be aware of a particular—in this case, the lack of consciousness—is to be conscious.[3] Such "unconsciousness" negates itself.

My Zen teacher once responded to supposed firsthand accounts of life after death by noting that "these people didn't actually die." Like so, while we can have firsthand accounts of the lack of conceptual thought—though only after the fact—we cannot have genuine firsthand accounts of the lack of Mind.

Mind is of a fundamentally different nature than the material world, or even the world of language. It's not merely relative. It can't be merely relative. It's of a different order, we might say. While Mind is Absolute Reality, language and the material world are mere conceptual manifestations resulting from the functioning of consciousness—which is an aspect of Mind.[4] Given this perspective—which does not rest upon any metaphysical speculation, but upon direct experience alone—neither language nor matter can be relied upon to account for either Mind or consciousness.

Furthermore, by acknowledging the primacy of Mind over matter, we'll discover that the historically intractable problems we've had with consciousness and reality are quickly surmounted, for we find that suddenly we're in no need whatsoever to account for or explain direct experience. We will find, however, that we must exchange our most tacit assumption about reality—namely that a thing is what it is—for one that is inconceivable. That is, direct experience reveals no substantiality at all.

In *perceiving this*, we'll also discover that we've exchanged a painful, senseless, meaningless world for a world that (apart from its inconceivable base) transcends meaning as we typically conceive of it.

Once we lose our habitual fixation on the objects (and subject) of consciousness—i.e., once we free ourselves from the trap of bogus ego consciousness—Reality is *seen* to be as much a seamless, boundless Whole as it is a collection of discrete parts. At this point, consciousness becomes easier to account for. **Consciousness can then be seen to be that which divides what is otherwise a seamless Whole.** In other words, consciousness is the conceiving (the making) of parts, or mind-objects, out of Wholeness.

It's the function of consciousness to divide subject from object—that is, to create parts or fragments out of what is otherwise a whole. Therefore the "parts"—the physical and mental objects of consciousness, i.e., concepts—are merely **appearances** resulting from the working of consciousness.

Our most grave, albeit our most common error, is to take these objects for Reality.

CONSCIOUSNESS, CONTRADICTION, AND CREATION

If we assume Reality to be fundamentally a seamless, Absolute Whole rather than a collection of entities, our commonsense view of things changes considerably. For starters, we must also assume that, since conscious awareness is obviously present in all conceptual experience (if it weren't, no such experience could be had), this **seamless** Wholeness would not merely have to **contain** conscious awareness, but would actually **be** conscious awareness. In other words, if Reality were a seamless Whole, that Wholeness **is** necessarily Awareness Itself—which is immediate and present in all experience.

Roger Penrose, however, in *The Emperor's New Mind*, noted that "To be conscious, I seem to have to be conscious **of** something."[5] But in stating this, Penrose, like Jaynes—and, indeed, like most of us—makes the classic Cartesian mistake of positing an absolute ("I") in his definition of the experience of consciousness. As a result of making this supposition—this concept—he, like most of us, misses what's otherwise immediate and unconditionally evident. And with the assumption of a self, it's not possible to actually *see* what conscious experience is.

To clarify this point, let's consider that famous question posed by Bertrand Russell (as well as by many others). Russell said that for him the great mystery was why there is something rather than nothing.[6] This question reveals the very ground of our commonsense belief upon which all our science has been built. There is, however, an enormous assumption being made here. As it was for Huai-jang (see page 13), to pose this question is to assume that Reality (which we habitually confuse with our objects of consciousness) actually **is** something, and that our objects (especially our physical objects) of consciousness actually **are** "somethings" as well.

But in making this assumption, we get into trouble with paradox. Physicists use terms like "ambiguous" or "indefinite," or in some way stress the contradictory nature of the world revealed in their quantum experiments. As we have seen, it's difficult to say much about a photon when it's not being measured (i.e., observed). In fact, the standard interpretation of quantum theory would have us focus on what is purely perceived—i.e., on maintaining an objectless Awareness—rather than assuming the photon to be something in particular.[7]

The standard (Copenhagen) interpretation of quantum theory, as espoused by physicist Niels Bohr, would not have us presume that a quantum system actually **is** something **before** a measurement is taken. But even **after** a measurement is taken (i.e., **after** we conceive the photon to be "there"), we run into problems. If we presume that a photon actually **is** something (as **opposed** to *nothing*)—just as we tacitly assume our common, everyday objects actually **are** something as we look at them—then ambiguity, and hence paradox, appear.

And, indeed, this paradox doesn't confine itself to the quantum world. Consider this curious ambiguity that results from conceptual thought: We move a cup from position A to position B. That a cup **can** actually move from A to B seems obvious, and is our commonsense assumption. But how can a single commonsense object **remain the same—remain itself—and yet move** through time and space? How can something endure, persist, abide, retain its identity—and yet change? Once the cup has moved from A to B, it's no longer the same object; yet we conceive of the cup at A and the cup at B as being identical.

We've already discussed a temporal version of this paradox in our considerations regarding our baby pictures. We want to think there's something—"I"—that doesn't change, that remains "I," even though we cannot point to any experienced thing that doesn't change. This is not unlike our persistence in seeing tapered tiers in the café-wall illusion (page 28), even after we notice that the defining lines of the tiers are parallel. We tend to hold to our commonsense notion even after we see that it entails a contradiction. And just as we tacitly believe a thing can persist and yet change, so too we commonly believe that a thing can move through space and time and yet remain unchanged.

We commonly believe that objects of our conscious awareness actually **are** something. That they're Real. That they persist as they are and yet change. But how can any such thing actually **be**? Contradiction characterizes the nature of all our objects of consciousness. If we only attend carefully, we can *see* that contradiction is the nature of **conceptual** reality. With bare attention alone, we can *see* that it is conceptual reality that cannot be.

WHAT IS CONSCIOUSNESS?

In Part I, I observed that we try to understand things in terms of their essence, yet it seems that we can understand things only in terms of their function or relationship to other things. And so it is with "things." "Thingness," however, is what consciousness **does**—that is, consciousness is the source of things and ideas. It breaks down Reality into pieces. It's the splitting of the Whole into conceived objects.

This is the functioning of consciousness: a conceptual awareness erupts out of Wholeness, thus splitting Wholeness; and the first thing that gets split off is: "Here **I** am, over here"—that is, consciousness is the conceiving of a subject along with its object. It's the conceptualization of a self. But more than that, it's the conceiving of a self that necessarily sees itself as being opposed to everything else. Expressed in a different way, consciousness is the spontaneous creation of a bogus self-consciousness.

Consciousness creates the appearance of an "other" that is set apart from "me." Thus the fragmentary, particularistic, commonsense point of view emerges from what is otherwise a seamless, boundless Whole. This, then, is the emergence of fragmentary consciousness, the origin of the fragmented mind that, in seeing itself opposed to "other," enters into perpetual conflict as it attempts to maintain what cannot be found in Reality—a self.

Consciousness then continues to function, dividing and redividing self from other, again and again, into finer and finer distinctions, more and more fragments. Thus the whole mental universe erupts into existence in a sort of "Big Bang." In other words, through the working of consciousness, the illusion of existence is born. Yet at every level, the objects of consciousness—including the subject, "I"—remain empty of any substance of their own intrinsic, separate, being.

THE MEASUREMENT PROBLEM

Nick Herbert noted in *Quantum Reality* that "It is fair to say that if we could say what actually goes on in a **measurement,** we would know what physical reality was all about."[8]

Why such high regard for what constitutes a measurement?

There's something about the act of taking a measurement that gives us the idea that there exists some fundamental difference between ordinary objects (like people and coffee cups) and quantum objects (like electrons and photons). But what is that something?

Ordinary objects, at least according to our commonsense view, seem to innately possess certain definite attributes such as, say, position and momentum. But these attributes—called "dynamic attributes" as opposed to "static attributes" (such as, say, mass or charge)—cannot be attached to quantum objects without qualification. The dynamic attributes of quantum objects arise only within a quantum object's "measurement context," which links the object to the rest of the universe, including any observer or measuring device. In other words, a quantum object's dynamic attributes are **contextual.** That is, it will exhibit different attributes depending on how we measure it. **Its dynamic attributes, in other words, are jointly shared by the object and the measuring device.** (Ultimately, of course, the measuring device is conscious awareness. For simplicity, however, let's say the measuring device is a conscious subject, a person.) Take away the measuring device and the "object" literally does not possess dynamic attributes—i.e., the photon isn't anywhere and has no motion (or lack of motion) when no one is looking. As Herbert put it, "We cannot picture such a state of being, but nature seems to have no trouble producing such entities. Indeed, such entities are all this world is made of."[9]

Of course we can't picture it. That's the point. It's not being measured—i.e., it's not registering in consciousness; it's not being conceived. Nevertheless, we perceive.

Herbert points out that John von Neumann, one of the twentieth century's giants in mathematics, showed that if we "...assume that electrons [or photons] are ordinary objects or are constructed of ordinary objects—entities with innate dynamic attributes—then the behavior of these objects must contradict the predictions of quantum theory."[10]

So why not reject quantum theory? Because quantum theory is the most successful theory in all of science. (Even Einstein couldn't defeat it, though he tried for many years.) Yet accepting quantum theory forces us to embrace ontologies that common sense finds absurd. Futhermore, Herbert noted,

> if you assume that electrons possess contextual attributes that stem from ordinary objects inaccessible to measurement but whose innate attributes combine "in a reasonable way" to simulate the electron's measurement-dependent behavior, then these entities likewise must violate quantum theory's predictions. Thus, according to the quantum bible, **electrons cannot be ordinary objects, nor can they be constructed of (presently unobservable) ordinary objects.** From the mathematical form alone, von Neumann proved that quantum theory is incompatible with the real existence of entities that possess attributes of their own.[11]

In other words, quantum theory is incompatible with our normal way of seeing things as substantial, definite, and real.

So, again, why don't we dump quantum theory? Because everything we have observed demonstrates the accuracy of quantum theory.

Could it be that it's our normal way of looking at things that is inaccurate? I am suggesting here that this is precisely the case.

According to von Neumann (and Herbert), electrons and other quantum objects cannot be ordinary objects, since they certainly do not—and cannot—behave as our ordinary objects of consciousness appear to behave. Yet every one of our "ordinary" physical objects (including our bodies) is made up of nothing but these extraordinary quantum objects.

I wish to suggest here that there is in fact no difference between "ordinary" and quantum objects—that **all** objects behave as quantum objects do. Everyday objects only **appear** to behave differently, and this appearance is the result of our rejecting direct perception (what I have called bare attention or *just seeing)* in favor of concepts. Despite our habitual denials, we actually **do** perceive "ordinary objects" (though "they" are not actually objects) just as we do any quantum event—but we don't attend to bare perception alone. Instead we overlay perception with conceptual thought, thus dividing the ordinary from the extraordinary.

In short, the distinction between ordinary objects and quantum objects is an error that results from our habit of packaging direct perception into concepts. If we would attend to perception alone, we would find that the world doesn't actually fit into concepts without yielding contradictions.

As John Casti wrote in his *Paradigms Lost*, "The paradox of the quantum realm is that although common sense dictates that the universe exists 'out there' independent of acts of observation, the universe does not actually seem to exist 'out there' independent of acts of observation."[12] But we're uncomfortable with such objectless Knowledge. We want a Real World, to be sure, but we'd prefer to have it with handles on it. We think that there's no other possible way we can "get it," in fact. Yet, as we can *see*, if only we'd *look*, whatever we "get" is never It. It's never the Real Thing we long to have.

The problem with our commonsense view is that it would have us insist upon a break between our ordinary view (the classical or Newtonian view of the world) and that of the new physics of mind-boggling quantum objects. But, though it seems to common sense that there's a vast difference between the quantum world and our everyday world, no such break is discernible. **Scientists have yet to find any evidence that our everyday world behaves any differently from the world of quantum reality.** As physicist Henry Stapp points out, "the ontology extends in an unbroken way [from the microscopic] to the macroscopic level."[13]

THE UNMEASURED SOLUTION

What we call "measurement," then, is what occurs when consciousness conceives an object and frames it in solid attributes. And from there we begin to analyze and synthesize. Our conceptual experience of this process is to have the sense that we've captured or recorded the essence of *something*. This "something," which was once dynamic, condenses into a concrete, conceptualized reality, where it now appears as though frozen in time and space. We commonly take the result of this process to mean that we've arrived at something "substantial" (as opposed to nothing, as Bertrand Russell might say). But what consciousness has actually done is merely form an abstraction and generate a mental object, a concept.

As we fix on our objects, we carry ourselves deeper into a "this means that" way of thinking and seeing. "This means that" is a way of seeing each thing as being separated out from the Whole and set apart from its "other." Left unchecked, this process of measurement and discrimination ultimately veils Awareness.

An example of this is reflected in the maxim "the best cooks don't measure." Good cooks trust perception, unlike ordinary cooks. These cooks don't **need** to measure—perception and experience are sufficient. They rely on wordless perception instead of measuring.

Whatever it is that makes a good cook, it has something to do with the cook's quality of mind. The mere acquiring of know-how isn't enough. A good cook gains practical knowledge, to be sure, but good cooks have also learned to merge with their objects. We often refer to such a process as inspiration. A good cook has a feel of the cheese, the butter, the eggs, the spices, whatever—they understand the life of this stuff, and, indeed, are merged with it. Their consciousness has loosened in such a way that they don't have to rely on a recipe. They **know** (even bodily) how to make the perfect soufflé. Though outwardly it may appear that they just throw the ingredients together, when they have finished—"voilà!"—out comes a masterpiece.

In the hands of a great cook (or musician, or carpenter, or mother), the quality of life thus rises to a high level.

WHAT IS MEASUREMENT?

Measurement is an obsession with the objects of consciousness. To measure—that is, to form a concept—is to not *just see*. Through measuring, we become less and less able to realize *what's* going on, and we lock ourselves more and more into conception instead of perception. This process occurs with extreme rapidity, and without reflection it becomes habitual.

We don't usually understand the act of measuring in this way. Rather, we tend to think of measurement as a way to obtain more information about our objects, so that we may better make use of them. We assume that such information is needed to make decisions about how we can further improve our lot. The more information we have, the greater our capacity will be to make wise and prudent decisions—or so we believe.

But this is a very questionable proposition. Indeed, gaining information through measurement is precisely what **limits** perception. One may have gained conceptual knowledge, but at the cost of having diminished the ability to directly *see what's* going on. To gain information is to conceptually remove ourselves, both temporally and spatially, from *what's* going on. We distance ourselves from Reality, as it were. (But, of course, it's impossible to deal with Reality at a distance, for that is already a violation of Reality.) To gain information is merely to sink ourselves deeper into a conceptual reality, and thus we place ourselves out of touch with *what's* going on. We gain information at the expense of true Knowledge, or Wisdom. And we don't even realize what we're doing.

The point that concerns us here is that more information does not make us any wiser or more capable of making sane decisions. And, though we live in an information age of supercomputers that can complete ten quadrillion operations per second, we still can't seem to get enough information to help us figure out what we're doing—or even what we're doing wrong.

Many of us are beginning to feel that the human world is fast approaching a crisis. With all our conventional knowledge—our ordinary, trivial, commonsense, conceptual knowledge, which is mere access to information—and with our fast communications, we only seem to dig ourselves in deeper. And the evidence for this has only increased since I published the first version of this book nearly two decades ago.

THE MEASUREMENT TRAP

I've loved classical music all my life. As a child, my greatest ambition was to be a composer. All through elementary and high school I collected records, and I listened by the hour to fine music on a small, aging portable phonograph. I loved doing this, and I didn't know that the phonograph was of poor quality.

Then in college, I purchased a stereo. I thought it was very good, but shortly after I bought it, I met a fellow who informed me that my nifty stereo was nothing to brag about. My new friend was familiar with electronic gadgets of all kinds. Soon he had me listening to my speakers and to my turntable. We analyzed my amplifier and my tone arm, and all the while, as he drew my attention to the shortcomings of each item, he suggested other models that might better suit me. We went over my set in such detail that I began to realize what a truly wretched machine it was. I deeply regretted that I hadn't met this guy a little earlier. He could have saved me from the misery of knowing I had purchased a lousy stereo.

He took me to see a friend of his who ran a store that sold stereo equipment. They had all the latest stuff. They hooked up the top-of-the-line amplifier to the best speakers and turntable, and put on a "full frequency range recording" of *Also Sprach Zarathustra* by Richard Strauss. As the stylus floated down to meet the revolving platter, the dealer turned to me and said, "Now here's a turntable, my friend!"

First the organ appeared in the lowest register, sounding the very bottom of audibility. The sonorities seemed to come from nowhere—or, rather, from inside my head. It was as if the room was breathing in and out with the slow resonance of the organ. The opening chord seemed to hang suspended in my mind until the orchestra swelled behind it, sweeping the sound upward, seemingly beyond the highest reaches of the ear. As that magnificent music surged, it was as though Zoroaster himself had emerged from his cave to face the sun within that very room.

My friends shouted to me above the tumult, "Listen to those cymbals! Sounds great, doesn't it?" And I exclaimed, "Oh yeah!"

I didn't like listening to my stereo after that. I became obsessed instead. After I got out of school, I earned enough money to buy the best stereo I could find. Eventually I was assured by my friends that I would have to spend thousands more to make any more improvements, so I figured I had at last acquired a good stereo.

But the funny thing was, I stopped listening to my music. I could no longer find anything in my vast collection that suited me. Within a year I got rid of it all—both my wondrous stereo system and my record collection.

I listened to music only when I happened to catch something on the radio every now and then. Oddly, I noticed that when it came over the radio it was quite all right. How different it was from when I sat before those shelves of records trying to settle on what I wanted to hear. And how at ease I was without having to pick and choose.

Over the next several years I rarely listened to music. After a long period of no concerts or records, I happened to drop in on an old friend one evening. As I arrived, he was about to listen to some music—Sibelius's seventh symphony. I had not heard it in years.

Without a word we sat and listened.

My friend was a very simple man. His phonograph was old and well-used. It was one of those old-fashioned suitcase-type phonographs—I don't even think it was a stereo.

I don't think I ever enjoyed Sibelius more.

It was perfect because I was no longer trying to measure, trying to package anything. I wasn't listening to the speakers or the turntable; I was listening to the music. I didn't care about how well the speakers or the performers were doing. That wasn't my business. That was the business of the musicians and the engineers. My business was to *just listen*, and to enjoy. If I had concerned myself with measurement, with grasping the moment, I would not have had the chance to actually *hear* the music. I would have only succeeded in disturbing myself.

My business was only to *just listen.* All else was out of my hands.

ON NOT BECOMING OBSESSED

When we concern ourselves with measurement, with getting a handle on things, we engage in an attempt to make *this* into "that." Our desire to change *this,* however, means that we have overlooked the possibility that, but for our meddling, *this* (whatever the immediate object of consciousness happens to be) is already well situated and quite clearly *knowable* for what it is.

In making *this* into "that," we often become confused and lose sight of what our intent was in the first place—yet we press on, oblivious to what we're doing.

Every morning I walk around the small lake near my home. One day I happened to look at my watch and noted that it took me 55 minutes to walk around the lake. The next day I noted the time again, but this time I found that it took me only 51 minutes. I had shaved four minutes from the previous day. And I thought, "I wonder how fast I can do this?"

Pretty soon I was down to 48 minutes, then 47 minutes and so many seconds. Soon I was really timing myself in earnest—right down to the second. I had a few setbacks from time to time, but then I thought, "I wonder if I can walk it in 45 minutes?" That became my goal. For the next few days I would not be distracted, not by geese flying overhead, not by the charming old man and his little dog, not by anything.

Finally, the day came when I thought I was really going to make my goal. Everything had gone well on the walk—there was no driving wind to slow me down, no distractions of any kind. I knew I was making good time.

I had only another 200 yards to my finish line when I came upon a large flock of geese. They stretched across the path and covered the lawn from the lake to the bushes on the hillside. Nothing but geese, everywhere.

I wanted to stop and watch them. I wanted to simply enjoy the beauty of the morning. But I thought, "No! I can break my record if I just keep going!" But the geese were in my way; they were going to prevent me from meeting my goal. I noticed I was feeling an odd mixture of delight (in the geese) and anger (at being barred from meeting my goal).

It was then that I caught myself. I realized that I had become so absorbed in my **idea** of what I wanted to accomplish that I had tuned out all other possibility. I was, in fact, ignoring *what* was going on. I was not only ignoring the beauty of the moment, but I was also not giving attention to the anger that was rising in me. My anger was rising over nothing of any consequence, and yet I was about to let it grip me.

I no longer wear my watch on my morning walks.

Truth is never glimpsed by an obsessed mind. Truth is glimpsed only after we let go of our intent.

Through our obsession with measurement, with our objects of consciousness, we continually distract ourselves. We then suffer because this keeps us out of step with *what's* really going on. We become increasingly less aware of our growing frustration.

It's easy to do this (and at the same time lose sight of what we're doing) when we measure.

All we can ever measure is quantity. I was measuring the quantity of time, and in so doing I wasn't paying full attention to the quality of time. I was measuring something that was far less Real than the actual quality of the moment. Through measurement we exchange, in effect, a higher order of Reality for a lower one. We exchange Reality for a representation of Reality—and a poor representation at that.

THE CONVERSION OF CONSCIOUSNESS INTO "THING"

The more we lock ourselves into thought constructs, the more cluttered we become with ideas and beliefs, the more fixed we become on measurement, and the more fragmented our conscious awareness becomes.

When we become obsessed with measurement, we lose sight of immediate Awareness, of direct experience of Reality. Without appreciating such Awareness, we develop an attitude of, "By George, let's get things done! Let's work hard to make life better! We must produce and improve and progress. We must build faster and better computers because we need to access more information, so that we may make the improvements this world really needs!" And we improve and improve until we develop a situation no one would ever choose to live in. And we run off to undisturbed places as soon as we have the chance—and then we "improve" **them!**

And all the while we do this, our conscious awareness—the quality of our life—is becoming ever more fragmented and confused. Then, in our more quiet moments, if we dare allow them to enter, we wonder what we're doing. And we can only shake our heads because we already *know* that things and ideas finally have no meaning.

In short, we become obsessed with objects of consciousness. And the more distinct our ideas and concepts appear to us, the more we latch on to them. Thus we become ever more enmeshed in a contradictory world—and less capable of either noticing that such is the case, or of dealing with it if we do.

THE AWARENESS OF WHOLENESS

It's possible, with bare attention, to *see* unity within diverse things. Such awareness is closer to perception than conception. The closer we get to pure Awareness, "things" are *seen* as interidentical with one another and with the Whole.

To demonstrate metaphorically how this occurs, let me slightly modify an image offered by the physicist David Bohm in his book, *Wholeness and the Implicate Order*. It runs as follows: suppose we have a rectangular fish tank with two glass sides perpendicular to each other. Inside

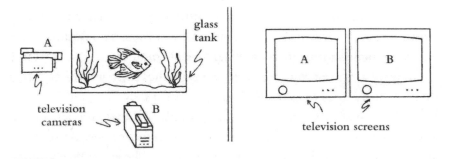

Figure 6–1. Bohm's Fish

the tank there is a single fish swimming around. Two TV cameras, at 90° to each other, view the fish through the two perpendicular glass walls. Finally, in another room, two side-by-side TV monitors display the two two-dimensional images being received by the two cameras.

Now, we should remember that most physicists believe there are more, perhaps many more, dimensions to reality than the mere three of our everyday experience. To get a feel for what it might mean to have more than three dimensions, let's step down a level, as we did in our balloon analogy on page 125, and imagine a two-dimensional being somehow looking at the two two-dimensional TV screens. These screens are displaying a single three-dimensional fish from two different viewpoints.

A 2–D being could not easily conceive of what existence would be like in three dimensions. Our two-dimensional friend, therefore, in viewing the two two-dimensional screens, would likely assume he's looking at two separate fish. And it would appear to this being that when the fish in screen A is facing straight toward him, the "other" fish in screen B appears in side view; and when the fish in screen A turns to the side, the fish in screen B turns its tail on him.

After a while, he might begin to realize that the movements of the "two" fish are coordinated, and from that he might deduce that they exist in some sort of causal connection. But though our friend may detect certain relationships between the two, he cannot easily conceive of how they might actually be connected, let alone how they might actually be one and the same fish.

We, on the other hand, being three-dimensional creatures, not only know there to be only one fish, but we know it on a level of awareness that is far richer than our two-dimensional friend can even imagine. Furthermore, we can understand why he believes there are two fish (multiplicity) rather than one (oneness).

Similarly, just as we might notice (though without any concept of how it might be so) that by harming others we ultimately do harm to ourselves, the two-dimensional observer may, without comprehending why, notice that what happens to one fish somehow affects the other.

He can see that they are "causally" connected somehow, though the thought that they are actually one and the same fish totally escapes him.

In like manner, we conceive that we exist in three dimensions, and we believe that we are separate and divided off from one another, as well as from the Whole. In Reality, however, it's possible to *see* that there's no separation, though in a way we cannot conceive. In other words, there's more about us that can be known than in ways we are able to conceptualize.

ORDER CANNOT BE ESTABLISHED

When we do not *see* the nature of Absolute Reality, we try to impose our own sense of order on things. We attempt to force our own ideas of order upon the Whole, upon Chaos. These attempts ultimately and inevitably create havoc.

Havoc, as I shall henceforth use the term, is precisely what we've commonly taken the word "chaos" to mean. But havoc is not at all like Chaos with a capital C. Chaos is of the Whole. As we have seen, it's stable, unique, creative, and generative. Havoc is none of these. Havoc results whenever we attempt to force Reality into nugget form. It's what we create when we try to establish order.

If you read about owls, you'll soon wonder how mice can survive. But if you study mice, you'll wonder how it is that an owl ever manages to catch one, or why we're not overrun with mice. If we see the whole picture, however, we can see how these two creatures balance their abilities; but if we try to set up that balance ourselves (i.e., by way of our intent), it will only lead to havoc, for this is not the place for us to let volition enter. It would be better if we let Chaos do the regulating and let ourselves simply observe and comply.

Volition's proper place is in directing us toward *seeing* the work of Chaos (i.e., the dynamics of the Whole), and in adjusting our living to fit the grand symbiosis, rather than striving to make everything fit the whims and fancies of small and contradictory propositions called "myself" or "us."

How we create havoc by imposing our sense of order upon natural systems is well illustrated in a story told by Joseph Wood Krutch in his book *Grand Canyon*. Krutch lived in the desert of the southwestern United States at a time when sheep ranchers were complaining that the puma (mountain lion) dined on mutton every now and then. The ranchers didn't feel it was right that they should have to sacrifice an occasional animal as a fee for the services of the puma. Indeed, the ranchers did not even recognize that the puma provided any service at all. They saw the puma only as a disturbance to the type of order they would seek to establish—namely the steady growth in profits from their sheep.

The ranchers pushed for unlimited hunting of the puma. They wanted to exterminate them. Eventually they got their way.

Within the very short period of thirty years, the extermination of the puma resulted in the "laying waste not only of hundreds of square miles of a once flourishing plateau clothed with many different shrubs and small trees but also, in places, serious damage to actual forests."[14] Once the puma were gone, the deer, now freed from a common predator, began to flourish. As their numbers mounted, they overbrowsed until, driven by near starvation, they began to "eat the shrubbery to the ground and desperately to gnaw the bark from dying trees." The resulting lack of vegetation, which had once protected the soil against erosion, led to still further disasters.

There was a small area, however, experimentally fenced in before the extermination of the puma. Krutch mentions that after thirty years it "does not seem to belong with the area outside." Outside the fence, "various species—some…among the handsomest [such as the Gambel oak and the beautiful cliff rose] have completely disappeared…where even the sagebrush is in a dying condition and the junipers have no branches for a deer to reach."[15] And many of the species that were not outright exterminated dwindled to numbers below which they could not rebuild their populations. Inside the fence, however, where the deer could not get at them, indigenous plant life flourished.

We can see this sort of shortsightedness repeated again and again. For example, we produced, used, and, to some extent, even made ourselves dependent upon the presence of chlorofluorocarbons (CFCs). All the while we did this, we unwittingly set the stage for the slow depletion of the ozone layer in our atmosphere. Year after year this went on undetected. We **thought,** of course, there would be no problem because such chemicals are relatively inert. But now we see that this very property actually intensifies the problem. We have awakened (to this one issue, anyway) to discover some of the unforeseen effects of this material in our atmosphere, and the disturbing realization that even after we stop dumping these gasses, the ozone layer will still continue to deplete.

The problem is our ignorance. We have to own up to it. We're the species that, in the New Mexican desert in 1945, wondered if our new experiment might ignite the atmosphere and incinerate the Earth. We discussed it and thought it unlikely, but we could not be certain. Yet we went ahead and exploded the first atomic bomb anyway.

When we set our sights on controlling change, just our sheer numbers can set an impressive edge to our intention. Consider this blurb from my local Nature Conservancy newsletter:

> Tropical forests are a stronghold of the Earth's biological diversity. Representing only 6% of the Earth's surface, they provide habitat for half of its wildlife species. For example, Costa Rica is only ¼ as large as Minnesota, but it has 58 times the number of species of trees, 3 times as many mammals, 8 times more reptiles, and at least 7 times as many butterflies. The American tropics is the wintering ground for at least 332 different birds that nest summers in the United States and Canada.
>
> About fifty acres of rainforest are being destroyed around the world every minute: 3,000 acres per hour, 27 million acres per year, an area approximately one half the size of Minnesota. With this destruction go many species of plants and animals including many not yet known to western science. The plants and animals [are] important in their own right.... The rainforests help stabilize the world climate patterns, substantially influencing the atmosphere of our planet.[16]

In all of these cases, our actions stem from our confused desire to make "improvements" in the way things already are, in the way Chaos arranged them. But it's just as clear that what we actually create is havoc, because we don't understand *what's* going on.

In Krutch's story we see that our humble desire was merely to control and improve. But no matter how altruistic our intent, there's always self-ishness in our desire. As long as we act out of our commonsense view of things, which forever holds us apart from the "other," our desire will always be centered on "me." And because our concern is with the part (ourselves) and not the Whole, we create havoc.

We—even those of us with the best of intentions—literally cannot conceive how everything is interconnected. We can only *see* (perceive) that it is so. In the words of Chief Seattle of the Suquamish Tribe,

> This we know, all things are connected like the blood which unites one family. All things are connected. Whatever befalls the earth, befalls the sons of the earth. Man did not weave the web of life; he is merely a strand in it. Whatever he does to the web, he does to himself.[17]

The ranchers saw the puma as different from the environment. (This is our commonsense view, or Proposition 1 of Nagarjuna's tetralemma—i.e., "a puma is a puma.") That's why they thought they could simply remove it and leave everything else, including themselves, unchanged. But is the puma different from the environment? Not really. Is it the same, then (Proposition 2)? No, not exactly—we see a puma, and it's in its environment, but we can talk and think about each separately. Is the puma, then, both the same **and** different from the environment (Proposition 3)? No, for how could we define the "puma"? Is it, then, neither the same nor different from the environment (Proposition 4)? No, for now we have even lost sight of what we're talking about. The fact is, the "two" are interrelated—indeed, they're not two.

So how **do** we explain what we experience as a puma and its environment? Don't forget $r + i$. A puma, as a concept—as an object of consciousness—is merely a unit unto itself (r). It's a "one." The environment is what we say is **not** the puma—i.e., we negate this "one." Yet in Reality, the "puma" is inextricably enmeshed with its environment and with the Whole ($r + i$), with Totality—and so is the conscious awareness that holds the puma as an object. Reality, therefore, is that the "puma" is implied by both itself and what it is not.

To put it another way, when we truly *see* a puma, we no longer merely conceive an object. Instead, we directly perceive an interrelationship, an interdependence, a dynamic interaction that is the entire cosmos itself. If we would truly *see* a puma, in other words, we must understand that the whole universe is necessarily involved in such perception.

Like the two-dimensional fellow struggling to grasp the true relationship between the "two" fish, it seems we have yet to accept that everything is one Reality that includes "everything else," like the living organs within a single body.

Wisdom is to realize that we cannot know the Whole—but knowledge of the Whole is precisely what we would need to govern the flow and flux of the Real World. We can't even predict the flow and flux of our little models of reality (such as the location and velocity of an electron—or, for that matter, next week's weather). How much more so are we unable to predict the dynamic of the Whole, which comes to reside within every mite and mote of being.

Yet, while we cannot know the Whole conceptually, we can and do *know* It. We *know* It directly and not as an idea. Unlike, say, a "banana squash," Truth isn't an abstraction. We all have the innate capacity to recognize the nature of Reality in the constancy of the patterns of Chaos, and in the immediacy of *this*.

seven

⎰(IMMEDIACY)⎱

*Space and Time! now I see it is true, what I
guess'd at,*
    What I guess'd when I loaf'd on the grass,
    What I guess'd while I lay alone in my bed,
    *And again as I walk'd the beach under the
paling stars of the morning.*
    *My ties and ballasts leave me, my elbows rest
in sea-gaps,*
    I skirt sierras, my palms cover continents,
    I am afoot with my vision.

—WALT WHITMAN

REALITY WITHOUT LOCALITY

It seems strange to suppose that when I lift the porcelain coffee cup
from my desk, bring it to my mouth, take a drink, and then set it down
again, my actions are intimately connected with everything else in the
physical universe. Yet we can infer this from modern physics—more

precisely from a theorem discovered in 1964 by physicist John Stewart Bell.[1] The implications of Bell's theorem suggest that when I drink from my cup, every atom on every star, even in the most distant galaxy, behaves in a way other than it would have had I not picked up the cup and taken a drink. What is more, according to Nick Herbert, Bell's theorem is derived from nothing more than a few facts and a bit of arithmetic. We might therefore accept this interpretation (or something much like it) with even more confidence than we can have in quantum theory itself. Wrote Herbert,

> Physics theories are not eternal. When quantum theory joins the ranks of phlogiston, caloric, and the luminiferous ether in the physics junkyard, Bell's theorem will still be valid. Because it's based on facts, Bell's theorem is here to stay.[2]

We don't ordinarily experience things in the way Bell's theorem suggests they are; yet on an intuitive level we can find instances where people **have** sensed this kind of reality from time to time. Francis Thompson, for example, wrote, "That thou canst not stir a flower / Without troubling of a star."[3] This now appears to be true not just according to our poetic intuition, but according to the "quantum facts." What you do now, at this very moment, changes stars in the most distant galaxy—at the very same moment you do it. Indeed, what you do changes **everything** in the universe in the same instant you do it.

If this is so—and Bell's theorem has allowed us to demonstrate clearly that it is—why isn't it more obvious?

Such reality ordinarily escapes our awareness because all such changes appear random. Herbert offers a hypothetical world to illustrate this point. He tells the story of Joe Green, who lives in a "**nonlocal** contextual world." What is a nonlocal contextual world? Says Herbert, "The essence of local interaction is direct contact."[4] He uses a gear train as an example. Motion is transmitted from one gear to the next in an unbroken chain. Remove a single gear and the transmission of motion stops. If there's not something there to mediate it, local changes do not get transmitted to distant places. The idea of locality is fundamental to all classical science, and clearly apparent to common experience.

Herbert points to voodoo as an example of a **non**local action. With voodoo, doing something **here** (sticking pins in this doll) effects a change **there** (injuring someone in a distant place) with no intervening communication or mediation. One object, event, or action can affect another that is far away and seemingly unconnected.

To most people of science, the idea of nonlocality has long been, as Newton put it, "so great an absurdity, that I believe no man, who has in philosophical matters a competent faculty for thinking, can ever fall into."[5] Bell's theorem, however, shows nonlocality to be a very integral part of Reality. (Bell himself was initially somewhat chagrined to have come upon nonlocality. In developing his theorem, he had hoped to prove just the opposite. A practice such as voodoo, however, is not supported by Bell's theorem since, as we shall see, intention or volition—i.e., communication—**cannot** be transmitted nonlocally.)

So here we have Joe Green in his nonlocal contextual world. As Herbert unfolds the story, Joe looks up into his sky and

> sees a rainbow made up of a glistening pattern of colored dots. Unlike the regular dots in a photographic halftone, Joe's rainbow's dots form a random array.
>
> On the other side of the same sun lies a counter-Earth, where Suzie Blue watches another rainbow in her counter-sky. Suzie's rainbow is likewise composed of a random array of colored dots. When Joe Green moves his chair, his rainbow moves too (a rainbow's position attribute is contextual, not innate), but Suzie's random array 200 million miles away instantly changes into a different (but equally random) array of colored dots. Suzie is not aware of this change—one random array looks pretty much like any other—but this change actually happens whether she notices it or not.
>
> The **phenomenon** in this hypothetical world, whether the rainbow moves or not, is completely local: Suzie's rainbow doesn't

move when Joe changes places. However, this world's **reality**—the array of little dots that make up both rainbows—is non-local: Suzie's dots change instantly whenever Joe moves his chair.

Such a non-local contextual world, in which stable rainbows are woven upon a faster-than-light fabric, is an example of the kind of world permitted by Bell's theorem. A universe that displays **local phenomena** built upon a **non-local reality** is the only sort of world consistent with known facts and Bell's proof. Superluminal rainbow world could be the kind of world we live in.[6]

Why would a physicist propose such an outlandish picture of Reality? Because Bell's theorem leads us to conclude that Reality is nonlocal. If it's not, we are faced with a paradox, a contradiction.

Let me lay out an extremely simplified version of the experimental proof for a nonlocal contextual world.[7]

If a photon is polarized (never mind what polarity in a photon means; even physicists don't know) at the same angle as the "transmission axis" in a sheet of polarizing film placed between the photon's emission source and a photon detector (see Figure 7–1, case A), there's a probability of 1 that the photon will be transmitted. If the angle of the transmission axis is at 90° to the polarity of the photon, the chance for the photon's transmission is zero (case B). At angles intermediate between 0° and 90°, the probability of transmission for any given photon will

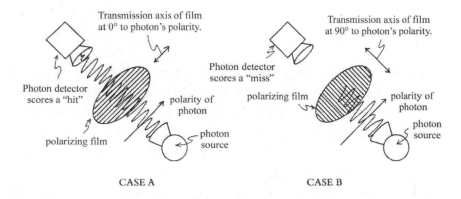

CASE A   CASE B

Figure 7–1

range between 1 and zero (specifically, the probability equals the square of the cosine of the specific angle involved).

According to Bell's theorem, if quantum theory is correct, there should be a higher correlation between the actions of two photons polarized in the same direction (or, as physicists say, "in the twin state"—which refers to more than just polarization, actually) than either common sense or classical physics would have us believe. Proof of this high correlation, and vindication of Bohr's interpretation of the quantum theory, have now been demonstrated in a number of experiments that show such high correlations.

The gist of these experiments goes as follows: two photons polarized in the same direction are ejected from a single source (C in Figure 7–2)

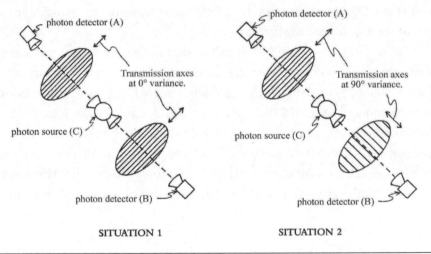

SITUATION 1   SITUATION 2

Figure 7–2

and sent off in opposite directions to two awaiting photon detectors (A and B) that are situated, relative to C, behind sheets of polarizing film.

In situation 1, where the transmission axes of both polarizing films are aligned in the same direction, there will be a 100 percent match between detectors A and B for any photon pair with the same polarization emitted by the source. In situation 2, where the transmission axes of the polarizing films are at 90° to each other, the correlation between what registers at A and at B will be a 100 percent miss.

If we mark the detection of a photon at either A or B with a 1, and the lack of such detection as a zero, then typical runs for these situations may look as follows:

```
situation 1  A     10011001110110101011000
situation 1  B     10011001110110101011000

situation 2  A     10011001110110101011000
situation 2  B     01100110001001010100111
```

The gist of Bell's proof has to do with the correlations between A and B when they are both set randomly at angles between 0° and 90°. For pairs of photons with identical polarizations, quantum theory predicts that the correlation depends **only on the relative angle** (let's call it Ø) between A and B and is independent of the actual settings at A and B. In other words, the correlation will be the same if A is set at 5° and B is set at 30° as it would be if A is set at 45° and B is set at 70°. This has been amply verified by experiment.

What this means is that in very long runs of photon pairs, for each set of angles we can predict the ratio of misses to matches. If Ø = 0°, there will be a 100 percent match. In other words, since $\cos 0° = 1$, there will be $(\cos)^2$ matches; that is, the results at A and B will match 100 percent of the time. On the other hand, at Ø = 90° there will be no matches, since $\cos 90° = 0$. Zero squared still equals zero; hence, no matches will occur.

These are simple facts. There's nothing mysterious or mind-boggling about what's been described so far.

But now let's look at such correlations in view of our commonsense assumption of locality. This assumption, unlike what the quantum theory predicts, says that changing the angle of the transmission axis of screen A will in no way effect the outcome (zeros or ones) found at detector B. This seems to be quite a reasonable assumption. It is, however, not borne out by experimental fact.

Consider: The transmission axes of the polarizing films at A and B are originally set at the same angle (they're both vertical). Say that by

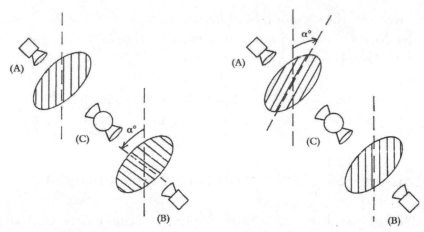

When A and B are misaligned by α°, one miss in four occurs:

case 1   case 2

A 10011001110110101010110000   A 11010001010111001011010100

B 10111000010010111011010010   B 10011001110110101010110000

  *   ** *   *   *   *   *   *   **   *

Figure 7–3

turning B α° in a counter-clock-wise direction (case 1 in Figure 7–3), we cause one mismatch of A and B out of every four trials.

On the other hand, if we only turned A's alignment by α° (in a clockwise direction this time), again this would only cause the same one mismatch out of four.

However, by turning them **both** by α° in opposite directions (thus creating a relative angle of 2α°), we might expect no more than two mismatches in four as we compare the results at A and B. Since this assumption overlooks the possibility that a change at A might coincide with (and thus, in effect, cancel) a change at B, the chances for mismatches may actually be a little less than two in four. If we take such possibilities into account, it would seem to follow that we can safely say that, over repeated trials, the sequence at A cannot score mismatches against the sequence at B by a value greater than two in four. The mismatches must number two in four **or less.**

We might expect something like this if we make the locality assumption, for it's surely what common sense would have us predict. The facts,

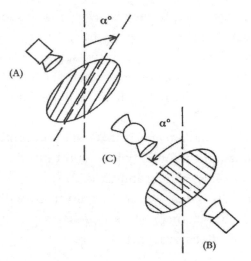

Since one in four mismatches occur when A and B are misaligned by α°, we should expect no more than two mismatches in four when they are misaligned by 2α°. For example:

A 1101000101011100101100100
B 1011100001001011101100010
　　★★ ★　　★　　★ ★★★　　　★★

Figure 7–4

however, are otherwise. When such experiments have been carried out, the number of mismatches of A against B in strings of paired photons average three mismatches in four, e.g.:

A　0101010111010100101101000
B　1011100001101011000100011
　★★★ ★★ ★★ ★★★★★★★ ★　★★★

This is precisely what quantum theory predicts: a very strong violation of the locality assumption. What it signifies is that by changing the alignment of the transmission axis at A, we instantly affect the outcome at B. Though the results at B are as random as before, we know that they must form a **different** pattern than they would have had we not made changes at A.

Bell's theorem has led us to the discovery that **no local reality can underlie the Real World.** In other words, though we conceive of a "here" and a "there," such conception is not supported by either bare attention to actual experience or by experimental results. The two are intimately related, and a change in one immediately creates a change in the other.

Again we see two that are not two. We see a multiplicity that is not multiple. But now our multiplicity that is not multiple, our two-not-two, is of the very fabric of time and space itself.

Linking the results of these locality experiments with quantum theory, and pointing out that there is "nothing that is not ultimately a quantum system," Herbert observes that

> **all systems that have once interacted at some time in the past**— not just twin-state photons—[are linked] into a single waveform whose remotest parts are joined in a manner unmediated, unmitigated and immediate. The mechanism for this instant connectedness is not some invisible field that stretches from one part to the next, but the fact that a bit of each part's "being" is lodged in the other. Each quon leaves some of its "phase" in the other's care, and this phase exchange connects them forever after. What phase entanglement really is we may never know, but Bell's theorem tells us that it is no limp mathematical fiction but a reality to be reckoned with.[8]

SOMETHING DOESN'T HIT THE MARK

Let's recap a bit, and take it slow. Our commonsense world is coming apart. When Bell made his discovery of interconnectedness (as this theorem has been called), he was trying to prove unequivocally the commonsense view that everything is **not** intimately connected. After all, this is how we normally see everything—how scientists see everything, anyway—and he surmised it would make everyone a lot more comfortable if we finally had some proof of it. In effect, Bell aimed to validate the common notion that "I" am separate from "you," and what I do with my cup has nothing to do with you, really (unless I fill it with coffee and offer it to you, or throw it at you, etc.). Yet experiments derived from his theorem ended up proving precisely the opposite.

The method Bell used in his attempt to prove this commonsense view was a simple arithmetical operation applied to a fact. Bell wound up showing that, starting with our commonsense assumption of a local reality, applying a bit of arithmetic to this assumption leads to a flagrant contradiction. Hence reality must be nonlocal.

In other words, Bell demonstrated mathematically that Reality is without locality. Much to his surprise and chagrin, he ended up revealing that *this* is all there is. *This*. No "that." Just *this*. And this doesn't make sense to us because we commonly see (i.e., conceptualize) "that" all the time.

Where did we go wrong?

In his "Background Essay," James Cushing noted that the word "reality" is defined as the "existence of an objective, observer-independent world."[9] Bell, Bohr, and Schrödinger, among others, have given us reasons to suspect that this definition of Reality is simply not correct. I suggest therefore that we remove some of the conceptual baggage (the inherent contradictions) of this definition, and instead define Reality as essentially "the way things are prior to conceptualization." By "Reality" we can mean *"what* is going on," and though *what* can't fit into concept, such can nevertheless be *seen* (i.e., perceived). Thus we have a more accurate definition of Reality.

To put it simply: we *know* (and can *see*) *what's* going on; we just can't conceptualize it. We need to learn to be at ease with inconceivability.

THE BODY FALLACY

Philosopher Linda Wessels, in her essay, "The Way the World Isn't: What the Bell Theorem Forces Us to Give Up," lays down the basic premises about quantum systems. But, like Descartes, she starts off by assuming something that is not given in direct experience. She begins with this basic assumption (she labels it "QS"): "the quantum systems measured in Bell's experiments can be treated as bodies." And what is a body? Says Wessels,

> A body is an object that is contained in a relatively well-defined, localized, spatial surface, thus has a well-defined spatial location, and in addition, remains distinguishable from other objects **even while its physical characteristics (including location and spatial surface) change.**[10]

This is a prime example of our commonsense view. I inserted the emphasis in the above quote, for it's here that we can notice how such a view overlooks the inherent contradiction that is otherwise so very obvious to bare attention.

Wessels continues,

> We commonly conceive of objects as bodies, and this conception of objects also underlies many scientific theories.[11]

Note Wessels' choice of words here: we **conceive** of objects as bodies, each neatly self-contained. **Conceive**, not perceive. But it's precisely conceptualization that is our problem. We confuse our concepts with Reality, with perception.

When Wessels lists examples of "bodies," she notes items such as tables, persons, clouds, and charged spheres. But when she comes to water molecules and subatomic particles, she notes that, "at least before quantum mechanics," these were thought of as bodies too.[12] But it seems now we don't know what to make of them. Indeed, Wessels herself suggests that her single most basic assumption about quantum systems, QS, may be incorrect. And indeed it is, as we shall *see*.

Quantum mechanics has finally brought us to the end of our rope as far as conceptualizing is concerned. For we have already noticed that none of our concepts are free of contradiction; that the gap between the quantum and the everyday world is not really there; and that we misapprehend what we actually "take in." (Indeed, the very fact that we would use a phrase like "take in" is indicative of our confusion.) We omit much, and not one bit of the world matches our commonsense view.

We don't dare abandon QS, however, says Wessels, for, "If we were to reject QS, how would we conceive of quantum systems?"[13] And that is the whole point: quantum systems can't be conceived at all! We can't see (i.e., conceive) live/dead cats. Rather, we can only *see* (i.e., perceive) *what's* going on, which we do (though most of us pay little attention).

Wessels notes that rejecting QS "would leave us with no notion of how to hang objective properties on quantum systems or how to model their interactions."[14] And that's quite right and fitting with direct experience, because quantum systems **don't have** objective properties. We don't find live/dead cats, for example. So if we reject QS, how would we conceive of QS? We won't; but we don't need to. "But," says Wessels, "if we give up the idea that quantum bodies (or fields or body-fields) have objective properties, then what is the point of retaining QS?"[15] Quite right. There's no point. The point is: **no** bodies have objective properties, quantum or otherwise. As we have seen, even our ordinary objects cannot be shown to reveal objective properties without also yielding blatant contradictions.

But then, observes Wessels, "Rejecting QS (and its field and body-field counterparts) leaves us with no intuitive model of quantum systems, and no idea of how to model their interactions...."[16] (In view of Nagarjuna's tetralemma, this would seem to apply to our ordinary, everyday objects—like pumas and coffee cups—as well.)

Does Wessels' observation mean, however, that we should instead, as an expedient, adopt a view that is demonstrably false? This begins to sound a great deal like the drunken man who searches for his keys in the wrong place because the light is better there.

In any event, Wessels' assertion simply isn't so. We **can** model quantum (and everyday) systems and objects. The simple description of $r + i$, the relative plus the inconceivable, gleaned from close and careful scrutiny of mental objects, does so perfectly well. It can also quell our itch to get a handle on things. That is, once we understand that our "objects" are mind objects only and are, therefore, without objective properties, we may relieve ourselves of the desire to slam our heads against the wall of conceptualizing "real" objects.

In her concluding remarks regarding what to give up in view of the implications of Bell's theorem, Wessels states that

> while the results of experimental and philosophical analysis of the Bell inequalities do require a significant departure from the way we standardly model physical objects, they have only minimal consequences for our conception of everyday objects and of most objects studied by science.[17]

Again, not so! And again, this is our problem. We do not easily give up QS because the invalidation of QS seems so disjoined from our everyday experience. But if we attend carefully to everyday experience, we can actually *see* that this is not the case.

Bell's Theorem shows us that all things are intimately connected spatially. But if we look closely, we can *see* that this also means we're intimately connected temporally. Everything that has ever happened or will happen influences what you are doing in *this* moment. What you do now, in this moment, effects a change in everything that has ever happened in the past and everything that will ever happen in the future.

We don't ordinarily think this is True, for it so strongly contradicts common sense. So what other signs are there that everything is so intimately and immediately connected in time?

LONG AGO AND FAR AWAY

Since 1963, scientists have discovered thousands of extremely distant quasi-stellar objects (QSO) in deep space. Quasars, as they are commonly called, appear as points of light, yet each quasar puts out more energy than a hundred supergiant galaxies. Quasars appear to reside beyond our galaxy and beyond our galactic family—indeed, even beyond the billions of other more distant galaxies that we can see in deep space. They are among the most distant objects we know about. Quasars reveal their great distance by their extreme "red shift," which, due to the expansion of space, gives them the appearance that they are receding from us at speeds up to 90 percent that of light. Some quasars appear to be more than ten billion light-years away. (A single light-year is about five trillion miles.) We are talking about objects that are at least ten billion years old—since we see them as they were when their light left them—and fifty sextillion (50,000,000,000,000,000,000,000) miles away.

The light coming from these quasars has been traveling through space for ten billion years. When we see a quasar—that is, when we just now detect the packet of light energy that left a distant quasar ten billion years ago—we detect light that has been traversing space at the rate of 186,000 miles per second for all that time without striking anything, until it finally reaches us on Earth.[18]

It's interesting to note that when the light we detect now left that quasar ten billion years ago, there wasn't any Earth. The Earth is only about five billion years old.

To give us some appreciation of how long five billion years is, let's condense the history of the Earth into a single year. The Earth, we'll say, formed on January 1st and it's now midnight on December 31st. All of human history—Christ, Buddha, Homer, everything that has ever been recorded from any ancient civilization—has all occurred within the past minute.

If we go back to our ancient ancestors, the Neanderthals, and even far before them, to Lucy and to the early hominids from whom we've evolved—this has all only occurred within the past hour of our condensed year.

If we go way back in time, way back before there existed anything that we would consider remotely human—back into what we commonly think of as the **very** distant past, such as the closing days of the age of the dinosaurs, we've only gone back to December 25th.

If we go back to mid-November, we're now at the time when macroscopic life first appeared on the Earth. Back even further, we find life itself beginning, and it is early spring. And if we go all the way back to January 1st, we find the nascent sun and its surrounding cloudy disk, from which planets are forming.

We've now gone back nearly five billion years, and that quasarian photon which will strike our twenty-first century eye would seem to be only halfway along its journey. To witness the event of the photon leaving the quasar, we would have to go back still another lifetime of the planet Earth.

HOW THIS MOMENT ALTERS THE PAST

Out of the many quasars known, some appear quite close together. After some experimenting, astronomers came to realize that a particular quasar identified as QSO 0957+561 appeared as a double image.

But how could one quasar appear in two places? The explanation for this phenomenon goes as follows: as light left the quasar and radiated out in all directions, some of it traveled past an intervening galaxy whose gravitational effect bent the quasar's light. In other words, the galaxy acted like a gravity lens, pulling the light around itself so that we observers on the other side see two images of the quasar on the opposite side (see Figure 7–5). The Earth is in a position to view its light as it passes around both sides of this intervening galaxy. While we detect two quasars, there's actually only one quasar positioned directly behind the galaxy.

Physicist John Wheeler constructed a "delayed choice" experiment that makes use of this galactic gravity lens and the double image of the quasar in the sky. By using conventional optics, we can focus and bring together the two light beams that have traveled around either side of the intervening galaxy and have them cross. If we measure the light—i.e., detect photons—at the position where the beams cross (position A in Figure 7–5), we get a wave interference pattern, signifying that each photon went around **both** sides of the galaxy. This is similar to what we found in the double slit experiment, where each photon makes use of two alternatives (i.e., both slits at once). By taking a measurement of each beam separately (at position B), however, we get a pair of diffrac-

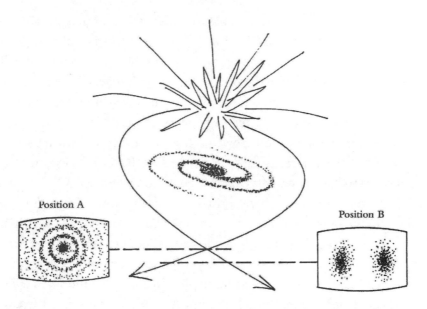

Figure 7–5. In Wheeler's "delayed choice" experiment, depending on whether we put our detecting device at A or B, we seem to be able to make the photon take both paths (option A, where wave interference shows up as a pattern of rings—the "Airy" pattern), or a single path (option B). Thus we seem to be able to determine in the present moment a situation that common sense would have us believe was settled ten billion years ago.

tion patterns, signifying—to commonsense, at least—that our photon either went around one side of the galaxy or the other. Again, this is like what we find in the double slit experiment when only one alternative (i.e., one slit) is available to the photons.

The photon we see *now*, however, left the quasar ten billion years ago. This would seem to be a fact. And which side of the intervening galaxy did it come around? This is a commonsense question, formed from the way we package our ideas in concepts. It would seem to common sense that the photon must have come around one side or the other—just as it seems a photon must go through one slit or the other in our double slit experiment. This seems only natural to common sense since, in conceiving a photon, we find it in a local context (i.e., it's here and not there; or, it went this way and not that way).

We know, however, that if we make two alternative paths available to our photon (i.e., we open both slits, or, in the delayed choice experiment, we place our measuring apparatus where the beams cross), it will take them both. But, unlike the double slit experiment, where the time duration from source to screen is so brief it doesn't seem to be a factor, we've now set up a situation where it appears, at least according to common sense, that our action now, in the twenty-first century—i.e., whether we place our measuring device at A or B—affects events that would seem to have been determined ten billion years ago.

This particular development, however, is only mind-boggling if we hold on to our most deeply set conviction of common sense—which in this case is expressed as our unquestioned belief in locality. We believe that a photon (or any object) **must** be either here or there. We cannot grasp (or, at least, accept) that a single photon came around **both** sides of a galaxy at once, even though this means that at one point the two "parts" (i.e., the two alternate paths, since photons only come in wholes and not in parts) of this photon were separated by hundreds of thousands of light-years. More mind-boggling still, our decision to put the measuring device at point A billions of years later made it so. All of this seems impossible to reconcile with common sense.

Locality means that our object—what registers in consciousness—exists *here* and nowhere else; but it also means that our object exists *now* and no time else. In other words, ten billion years ago is indeed very distant; and ten billion light-years away is indeed very, very far away as well. But locality shows only one aspect of our life and mind—the very ordinary *r* aspect that we all assume. With this understanding we conceive that we're all separate from each other. And we imagine that what I do has really nothing to do with you unless I either directly interact with you, or, at the very least, directly communicate through some medium. This is our assumption of locality. But as long as we hold it, we're out of step with Reality. And when we don't *see* how we're out of step, we suffer.

To put it in everyday terms, what you or I do right *here*, right *now* affects everything that ever was, is, or will be.

THE GRADED STREAM

When I was in college, one of my geology professors introduced me to the idea of a "graded stream." One day he walked into the classroom, went to the chalkboard, and, without saying a word, wrote down the definition of such a stream, as given by one J. Hoover Mackin. A graded stream is one

> in which, over a period of years, slope and channel characteristics are delicately adjusted to provide, with available discharge…just the velocity required for the transportation of the load supplied from the drainage basin…. Its diagnostic characteristic is that any change in any of the controlling factors will cause a displacement of the equilibrium in a direction that will tend to absorb the effect of the change.[19]

The graded stream is an image of dynamic balance and interconnectedness. As one part of the system undergoes change, all other parts of that system move in such a way as to acknowledge and accommodate that change, thus maintaining balance. It's a system that is in a state of endless becoming.

To illustrate how this dynamic appears in the physical world, let me give you the simple example my geology professor used. After the last great melting of the glaciers that covered North America thousands of years ago, there formed a Great Lake, called Glacial Lake Agassiz, which, at its height, covered most of Southern Manitoba and much of Ontario, Minnesota, and North Dakota. It was an enormous lake that drained through the Minnesota River and the Mississippi.

Lake Manitoba and Lake Winnipeg, in Canada, are present-day remnants of Lake Agassiz, which has been steadily shrinking. About ten thousand years ago, Lake Agassiz was reduced to a size where it no longer drained through the Minnesota. This resulted in a great reduction in the amount of water flowing through the Minnesota River. This, in turn, reduced the amount of flow in the Mississippi.

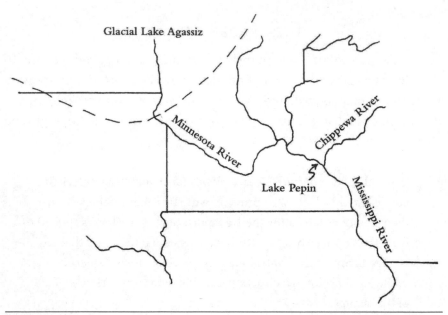

Figure 7–6

Now, downstream from where the Minnesota empties into the Mississippi, the Chippewa River flows into the Mississippi from the Wisconsin side. The Chippewa carries a lot of sediment, and when the amount of water flowing through the Mississippi was suddenly reduced for lack of effluent from Lake Agassiz, the Chippewa River simply dropped its load right at the point where it enters the Mississippi, thus forming a cataract. Behind this cataract, Lake Pepin was formed. You can visit Lake Pepin today; it's simply a wide place in the Mississippi River.

Eventually, if the Army Corps of Engineers leaves it alone, Lake Pepin will disappear, for the Mississippi is in the long, slow process of adjusting itself to meet the demands of the Chippewa River. In the slowed waters of Lake Pepin, the Mississippi drops its load of sediment. In time, this accumulating sediment will fill Lake Pepin up to the level of the cataract. Once that occurs, thousands of years from now, the river will then begin to down cut through the cataract until it reestablishes a new grade, which will be just steep enough to sufficiently increase the velocity of the flow, so that it can handle the load coming in from the Chippewa.

The graded stream is, in a limited sense, a self-regulating system. It's an image of how a whole continuously calls on all its parts to change in a coordinated, unified way, thus maintaining the dynamic equilibrium of the whole.

YOU ARE "THAT"

There are two aspects of our existence. One is called "this is it"—the *this,* the *"something,"* or *r* aspect. It's here that we exist as separate entities, in a particular place, at a certain time.

But we must not forget that there is another aspect called "what is it?"—the *what,* the *"nothing,"* or *i* aspect. The two aspects are interrelated and interpenetrated; they are like a seiche, the back-and-forth movement of liquid in a basin. A seiche constantly spills out of itself and into its "other," only to slosh back. The *r* and *i* aspects are also like a graded stream, where as soon as something in the system changes, everything else in the system—which involves stars and galaxies, as Bell's Theorem demonstrates—begins to move to counter the effect of the change.

So when we ask, "what is it?" we can only point to "here it is." "*This,*" is all we can say. It—whatever "it" happens to be—constantly exchanges its identity with every other thing. This is how we live. We live in a Reality that is like music, like a graded stream, or like the sloshing of liquid within a basin. We "exist" not in **being** but in becoming—and in fading away.

Within one aspect of our lives—the common, bounded, *this* aspect—we each have separate identities. But we must also accept that "other" aspect that reveals no boundary. Given this other aspect, each object and each person is intimately connected (indeed, is interidentical) with everything that ever was or ever will be, no matter how distant it appears to be in space or time.

Once we realize this other aspect of Reality, we can *see* that there's something more to human life than mere phenomenal existence.

There's something vast, wonderful, and unbounded. There's a deep relationship, a grand symbiosis, an interidentity of Whole and part.

NOW

The problem of life and death is a big one for us. This is because—unlike other animals, supposedly—we're fully aware of our own inevitable death. We must all face the prospect of our passing out of existence. Let's consider this problem now, for it has everything to do with our conception of "now" and our ideas about before and after.

We have defined consciousness as the division of Reality or Truth. Consciousness is the division of what intrinsically has no division or parts, but otherwise remains a boundless, seamless Whole. We have thus far defined this division of Reality primarily in terms of space—as in "I am over here and you are over there." Consciousness splits the Whole, immediately creating an ego—an identity—that then sees all other things in opposition to it.

We'll now consider this mental function—i.e., consciousness—again, but this time, rather than in spatial terms, we'll look at the splitting of Reality in a temporal sense. In other words, we'll look at "now."

"Now," if we look at it carefully, suggests that there's no future. The future is something that is not Real because it's something that isn't *Here*. The future is what hasn't happened yet. Still we think, "Well, okay. But this doesn't mean it isn't Real; it just means that it hasn't gotten here yet."

But, you see, the future is incapable of **ever** "getting here" because all we **ever** experience is *Now*. We commonly think that time is flowing and moving on, and that eventually the future will get *here*, but if we look at this carefully, we can *see* that we never live in the future. We can daydream about the future, but we can't live there. The future is forever not here. Though we think, "But I will someday arrive in the future," it's clear that "I" will never arrive in the future for, as "I" am not that person in my baby picture, so too, whoever that person is that we believe we are referring to as "arriving in the future" is surely not this "I" appearing *Now*. Each of us changes in every moment—in fact, we're nothing but change. Even as common sense would have it, each of us will appear subtly or profoundly different in the future than we do at present.

Furthermore, the future of our commonsense reckoning isn't really The Future. What we think of as The Future doesn't exist *Now*. When we actually experience "the future," it's *Now*. When the future "arrives," as common sense would have us believe it does, it's just as much *Now* as every other moment we conceptualize.

Likewise we cannot grasp the past, for the past, as we imagine it, isn't Real—it isn't *Now*. We have our **memories,** to be sure, and we commonly think the past **is** Real, but, if it's Real— if it exists—where is it? Where is, say, the event we call "Columbus-discovers-America"? It's not *Here*. Search for it as we might, we won't find this event. But, of course, we don't expect to find it because, as we say, it's in the past, and, as we all *know*, the past doesn't exist *Now*.

Even the event we remember as "turning-to-this-page" can't be found. It doesn't exist *Now*. There's no Real event corresponding to the phrase "turning-to-this-page." It's merely a relative event, a concept, tucked away in memory and probably soon to be forgotten. It's not *Now*. Not as you read this. Not in Reality. So, where **is** "turning-to-this-page?" How did we get *Here* if it isn't Real? How did we get to this dynamic event that is *Now?*

The Truth is, we didn't **get** *Here*. *Now* is where we've **always** been.

Our common sense may have us think that what's being discussed here is nonsense—or, at least, just a play with words that points to nothing more than vague abstractions having no connection with the Real World. But in this commonsense assessment, we're completely turned around. It's because we habitually react in this way that we remain confused and never find our way to Truth.

On the contrary, what's being indicated here is immediate and Real. It's our conceptual reality, where we spend virtually all our intellectual life, which is utterly abstract. And, with common sense, we habitually confuse our abstractions with Reality.

What we commonly call "now," unlike the past and the future, is Absolute and not relative at all. It's *Now*. *Now* holds both past and future and "takes place" outside of time. It's for this reason that we can't hold *Now* in concept. As we shall see, we can't hold *Now* in any way whatsoever.

According to our commonsense view and its locality assumption, we believe we can substantiate a past and a future. We believe they exist "out there." And because we believe they are "out there" in some substantial way, we reify this notion and start to build upon it. Soon we imagine such a thing as time travel. As Nick Herbert put it,

> If time is like space, then the past must literally still exist "back there" as surely as Moscow still exists even after I have left it. If the past still exists, then it makes sense to consider whether one could actually travel there....
>
> If we take the fourth dimension seriously, we must believe that past and future have always existed, and that human consciousness, for reasons we do not comprehend, perceives this "block universe" one moment at a time, giving rise to the illusion of a continually changing present.[20]

Herbert then quotes mathematical physicist Herman Weyl as he describes our limited "perception" (but what is actually a conception) of a four-dimensional space/time world:

> The objective world simply is; it does not happen. Only to the gaze of my consciousness, crawling upward along the life line of my body, does a section of this world come to life as a fleeting image which continuously changes in time.[21]

But these are not observations based on direct experience. Rather, these are reports founded on our commonsense notions, our concepts; thus they are necessarily fraught with contradictions.

We cannot travel to the past or future simply because we can never leave *Now*. Indeed, the past and the future are nothing but conceptual constructs. Such contradictions as time travel are not in the world itself, but in how we package the world in thought.

In our actual experience all there ever is is *Now*. Bare attention reveals this lucidly. At this very moment we live in *Now*. When we actually experience 5 p.m. on Sunday, we experience it as *Now*. This is, was, and always will be true of any point in time. There's no contradiction in *Now*. We can only find contradictions in how we conceive of (i.e., break up and separate *Now* into) time.

Roger Penrose was right on target when he noted that

> The temporal ordering that we "appear" to perceive is...something that we impose upon our perceptions in order to make sense of them in relation to the uniform forward time-progression of an external physical reality.[22]

NOW IS UNGRASPABLE

We cannot stop with the mere observation that past and future exceed our grasp. We must realize that even *Now* is ungraspable. Snap your fingers and try to hold the snap—but you cannot, for already it's past. It's not *Here*. It's not *Now*. It's already memory.

We can't even get our hands on the immediate *Now*, for it has no duration; even so, *Now* is where we always live.

To illustrate this point, let's consider the strange creature Amphibius, which philosopher James Cargile has thought up.

Amphibius, when we meet him, is a tadpole living in a bowl of water. We film him continuously for the next three weeks. At the end of three weeks, Amphibius is a frog. If our movie camera records twenty-four frames per second, at the end of the three-week period we would have about 43.5 million consecutive pictures of Amphibius. We then number the frames 1 to 43,500,000 in the very sequence in which they were shot. Frame 1 shows a picture of a tadpole; frame 43,500,000 shows a frog.

According to Cargile,

> there will be one moment when Amphibius is a frog, such that, an instant before, he was not....It is not being denied that, for the

young tadpole Amphibius it will be a long time until he is a frog…
growing can take lots of time. But acquiring properties does not.[23]

Why not?

If, as our common sense tells us, the creature in frame 1 is "just" a tadpole—that is, a clearly-defined entity of that name—and the creature in frame 43,500,000 is clearly and separately a frog—then, according to logic, there must be a frame showing a tadpole, which is immediately followed by a frame showing a frog.

This absurd premise is predicated on the "principle of the least number," which is a theorem of mathematical logic. This theorem states that in any given series 1 to n, if 1 (a tadpole, say) has some defining characteristic that is lacking in n (a frog, say), then there must be a "least number" of members to the series (say a million, or 37, or even 1 out of the 43,500,000 frames in the series) that do not have the defining characteristic in question.

But there can be no "acquiring of properties." Though this theorem accommodates our conceptual packaging, clearly it doesn't accommodate what we directly perceive.

As the Zen master Dogen Zenji put it, we do not see spring become summer. Spring is spring. Summer is summer. Spring, however, **is** spring precisely because it's also not *just* spring. And this, the *what* aspect, is just what common sense would have us overlook.

In other words, if we would avoid contradiction and confusion in our accounts of experience, we must account for more than just the mere r aspect (i.e., spring is spring) alone. We must not omit the i aspect. Spring is $r + i$. Spring can **only** be spring if we account for what it is not (e.g., summer) as an intrinsic part of its identity.

A purely logical account, on the other hand, as we see with Cargile's account of Amphibius, entails a contradiction. Logic, as long as it remains based in conceptual thought (e.g., as long as it assumes the law of identity, as Aristotelian logic does), must demand that a frog is **only** a frog. It makes no attempt to account for "frog" within the relative frame that would include, among all else, "tadpole."

In other words, spring can **only be** spring because it's also not spring (i.e., because it holds all that it is not as part of its identity).

Cargile attempts to account for his frog as though it were Absolute Frog—but he can't account for either the frog or the tadpole according to the *r* aspect alone—i.e., as though it were Absolute. This is precisely why his thinking leads us to this particular paradox. He's left out the *i* aspect, which is apparent to bare perception. Logic, because it's purely conceptual, can make no allowance for the *what* or *i* aspect of Reality in any account of experience. Indeed, by itself logic makes us oblivious to Reality.

Truth is beyond "before" and "after." *This* moment is beyond before and after. We never experience spring as summer; we never even experience spring becoming summer. We only experience *Now*.

LARGE AND DIMENSIONLESS NUMBERS

There's one more point about *Now* that I want to look at, but in order to do this we should step aside for a moment and consider a few facts that have recently come to light about large numbers. We don't ordinarily deal with large numbers in our daily life, and we're generally not very aware of how quickly numbers can grow and propagate. Nor do we appreciate just what significance large numbers may have for us.

To help us put large numbers in perspective, let's consider the difference between a million, a billion, and a trillion. Though the ratio between a trillion and a million is the same as that between a million and one, many people do not think of there being much difference between a million and a trillion. For many, if the national debt were a billion or a trillion dollars, they'd pass it off as a big number and let it go at that. (If you're one of these people, I recommend the book *Innumeracy* by John Allen Paulos [Hill and Wang, 1988].)

Suppose some omniscient being promised us that world peace would be established either after a period of one million seconds, or one billion seconds, or one trillion seconds from now. How long will we have to wait? If we only had to wait one million seconds for peace on Earth, we'd have it in less than two weeks. If we had to wait a billion seconds, those of us who are not too old now might have the chance to see peace commence within our lifetimes. We'd have to wait almost 32 years, though.

And if we had to wait a trillion seconds for world peace, then we should never hope to see it in our lifetimes. In fact, it might never come at all, for humankind would have to live with war for another thirty-two thousand years, and it's doubtful that we could survive either our ingenuity or our ignorance for that long.

To emphasize how we tend not to notice the speed and degree to which numbers can grow, let me recall a fairly well-known story of a king who was introduced to the game of chess. The king was so delighted with the game that he asked its bearer to name his price. "Anything," said the king, "just name it and it shall be yours."

"Sire, I have but a simple request," said the bearer. "The chessboard has sixty-four squares. I ask only that you give me one grain of rice for the first square, two grains for the second square, four grains for the third, eight for the forth, and so on to the last square upon the board."

"Such a modest request!" said the king, sending off his servant to fetch a sack of rice from the royal store.

As the story goes, the king soon had to send for more bags of rice, and then for cartloads. Before long his store was empty and he had not even approached the sum required to fulfill his debt. What the king did not realize, and indeed what most of us would not suspect, was that in doubling the amount at each square, very large numbers are quickly generated. The amount of rice the king owed totaled 18,446,744,073,709,551,615 grains. That is more than eighteen quintillion (eighteen million million million) grains of rice.

Such a large number is easier to grasp when it's expressed in some graphic way. For example, the king's granary, if it were 40 stories high, would have to cover 800 square miles to hold that much rice. That is, the rice would fill a square building 450 feet tall and more than 28 miles on a side. (A single cup of rice, incidentally, contains about 11,000 grains.)

Now I want to talk about a **really** big number—a number that is considerably larger than the number of those rice grains multiplied by itself several times over. Let's consider the number one hundred million million million million million million million million million million million million million. This number, written out in standard numerals, would consist of a 1 followed by eighty zeros. Scientists, however, use the simpler notation of 10^{80}, or ten to the eightieth power. This equals ten multiplied by itself seventy-nine times. To Sir Arthur Eddington, the physicist and astronomer, this number represented the approximate number of protons in the observable universe. We'll be returning to this number shortly.

Now let's consider what are called "dimensionless numbers." A dimensionless number is simply a number that does not express any particular unit of measurement. For example, 5 pounds is **not** a dimensionless number. Nor is 5 miles. But the number 5 **is** dimensionless, because 5 by itself does not designate any unit of measurement.

There are various ways in which units of measurement can be removed from dimensioned numbers. For example, if we take 10 pounds and divide it by 5 pounds (10 lbs./5 lbs.), we'll find that the pounds will cancel out and the number 10/5 reduces to 2. In other words, we get rid of the pounds and we end up with a dimensionless number. In this case we end up with simply the number 2, which signifies that 10 pounds is twice that of 5 pounds.

We can apply this type of operation to all sorts of natural phenomena. For example, when we divide the mass of a proton by the mass of a neutron, we end up with a dimensionless number, because the terms indicating mass cancel out. In this case we end up with a number that is very close to 1.

It so happens that when we perform mathematical operations upon the constants of nature in such a way that we cancel out the units of measurement, we usually end up with numbers that are very close to 1. There are a few exceptions to this rule, however. In these, instead of numbers near 1, we find enormous numbers.

We are now about to explore one of the great "mysteries" of the physical world. Scientists have been pondering this ostensible mystery for some time, and a great deal of philosophical speculation has arisen regarding it.

THE LARGE NUMBER HYPOTHESIS

It all began when Sir Arthur Eddington took several of nature's constants and combined them in such a way as to get rid of their dimensions. The constants he used were the speed of light, the mass of a proton, the universal gravitational constant, and Planck's constant. (Planck's constant is a universal constant (h) that gives the ratio of a quantum of radiant energy (E) to the frequency (v) of its source.[24]) These constants carry dimensioned terms that cancel out when put through a few simple arithmetic operations.

Why are these entities called "constants"? Basically, it is because they refer to qualities or factors—i.e., not things, but relationships—in the natural world that do not change. They remain constant while other values do not. For example, the mass of a proton (at "rest") does not change. The electric charge on an electron does not change. The speed of light does not change.

For example, let's briefly reconsider the speed of light. This is a constant because, regardless of who's looking at it, the speed of light in a vacuum is always found to be 186,000 miles per second. As we already noted in Chapter 3, no matter what our speed or direction of travel is relative to a source of light, all observers see light traveling at 186,000 mps. The speed of light never varies.

Eddington took the speed of light (c), multiplied it by Planck's constant (h), and divided that quantity by the gravity constant (G) multiplied by the mass of a proton squared (m_p^2). The reason he did this was simply to get rid of the units of measurement—in other words, he wanted to end up with just a plain, dimensionless number, a ratio. Eddington was somewhat intrigued when the number he came up with turned out to be rather "significant." The number he found was 10^{40}, or ten to the fortieth power:

$$hc/Gm_p^2 = 10^{40}$$

Eddington, as you'll recall, was aware of the significance of that larger number we looked at earlier—the number 10^{80}. And if we take 10^{40} and multiply it by itself, we end up with 10^{80}. He found his new discovery to be rather curious because, of all the possible numbers that lie between zero and 10^{80}, he ended up with 10^{40}! In other words, Eddington discovered that by arranging these four constants of nature so as to cancel out their units of measurement, lo and behold, we find the square root of his estimated number of protons in the observable universe.

This outcome intrigued him because there is no obvious reason why these constants should yield such a significant number. He thought that there must be some deep, underlying principle of nature at work here. Most physicists at the time thought it to be mere numerology. Even so, the odds of hitting such a number by chance are incredibly minute.

HOW OLD IS THE SKY?

It was several years later when physicist Paul Dirac discovered another startling "coincidence." Dirac made his discovery when he decided to measure the age of the universe. (As we've already seen, the Universe is literally timeless—there is only *Here* and *Now*. But we can't neglect the *r* aspect of the Universe, and Dirac wanted to take one measure of it.) But, rather than using our everyday units of measurement, Dirac chose a unit of time that was universal—or, we might say, constant.

Our standard units of time are not universal. Cosmologists currently estimate the "age" of the universe to be roughly 13.7 billion years.[25] But a year is merely the time it takes this one planet, out of the billions and billions of planets that exist in the universe, to go around its star, the sun. The unit of time we call a year, therefore, has virtually no significance, cosmologically speaking.

The same is true of the second. A second is merely a particular division of the day, which is in turn determined by the rotation of this planet. This unit of time also has no universal meaning.

Since Dirac wanted to use a fundamental unit of time, he needed to find one that depends only upon natural constants (such as the mass and charge of an electron, and the speed of light). The one he chose happened to be one that physicists associate with atomic and nuclear processes. Dirac's unit is the time it takes light to traverse a fundamental distance known as the "classical electron radius"[26] (which is merely a characteristic distance associated with nuclear processes; it should not be taken for the "true" radius of an electron, since the very concept is meaningless). This unit of time, since it's derived from natural constants alone, ought not to change from place to place, or from time to time. And it's a very short period of time indeed, because light travels very fast, and electrons (as we conceive them, at least) are very small. Dirac's time unit is about $1/10^{23}$ second, or 10^{-23} second.

When Dirac calculated the age of the universe in these fundamental units, he came up with 10^{40}! In other words, 13.7 billion years equals roughly 10^{40} of these fundamental units of time, which are derived from natural constants. Once again we hit the square root of 10^{80}.

Dirac also discovered that the ratio of the electrical force between a proton and an electron divided by the gravitational force between the same two particles **also** yields 10^{40}.

These are enormous numbers, we must remember. And, as Clifford Will observed, "of all the possible powers of 10 between 0 and 100, say, why should two such dimensionless numbers, arrived at by very different reasoning, come so close to each other?"[27]

This is surely quite interesting and, at the very least, very strange—unless something more than coincidence is involved here. Dirac suspected something. Eddington did, too. Indeed, many physicists are beginning to suspect that perhaps there really is some significance to these numbers.

But, in fact, Dirac has now uncovered an even greater "mystery," because it would seem we're no longer merely dealing with constants—that is, we're no longer dealing with things that don't change. Dirac, as we recall, brought the age of the universe into his calculations. But from our relative, conceptual viewpoint, the universe is getting older—it's not constant! It's changing all the time. Time marches on—at least so it appears to human consciousness. According to common sense, then, it would seem that 10^{40} is not destined to remain 10^{40} forever. Yet what an enormous coincidence it is that so many constant relationships converge on this number.

THE ANTHROPIC FALLACY

In an effort to explain these coincidences, some physicists have invoked what is called the "anthropic principle." The name is derived from the term "anthro," which means "man." It is so named because it would seem that this principle has everything to do with us. The anthropic principle states that the reason the number 10^{40} appears *Now* is because we're here to see it.

According to this principle, in the distant past, had human beings been there to look, we would not have found 10^{40} by manipulating these constants. But then, of course, we **weren't** there—we hadn't evolved yet. And in the distant future, when this ratio eventually rolls on to a noticeably changed value—and it will take some time for it to do so—we won't be here, because by then the stars will have given up their fire and there will not be any new ones coming along, etc. It seems, according to this principle, that only now, while there's 10^{40}, could conscious beings like us be around to find 10^{40}.

This sounds pretty good, except that it has a tautological ring to it. It seems to say little more than "things are the way they are because that's the way they are." It's a response that resembles our commonsense response to the law of identity: "a thing is what it is." Furthermore, it places human beings temporally at the center of the universe—in much the same way that, centuries ago, we thought of ourselves as occupying the spatial center of it all, simply because it appeared to us that the sun and all other heavenly bodies revolved around us.

The anthropic principle is our attempt to explain these numerical "coincidences" from our commonsense point of view, which tacitly assumes the primacy of matter over Mind. But much of human experience lies beyond the physical sphere. As we have already noted, our conception of matter—of physical reality—cannot be accounted for apart from conscious awareness. Furthermore, we have noted that matter cannot be explained through the use of any device within the material world, such as, say, time. To do so only leads us to "a thing is what it is."

Such attempts to explain the world through concepts are no more than the endless rearranging of furniture in a room. Yet, as we've seen, we must also take note of the room itself.

As we've also seen, conscious awareness cannot be constituted of atoms. And, just as we cannot remain solely within the material realm if we would arrive at some understanding of conscious awareness, so, too, we cannot remain solely within the physical realm if we would actually *see* physical phenomena for what they are. If we wish to *see* the significance of 10^{40}, we must draw in the *what* aspect, which is always lacking in our commonsense view.

Seeing is required to get to the bottom of things.

As we've noted, we would have to step outside the Whole to get a look at It. But we can't do this because, as we already *know*, the Whole is boundless in time as well as space. There isn't any Real Past. We can't get back to the time of the Big Bang, when, at $t = 0$, time, as well as space, erupted out of *nothing*. It's not *Here*. It's not *Now*.

In Reality, there can be no edge to what we call time, for the Whole of time is boundless—it's always found to be *Now*.

I wish, therefore, to suggest an alternate interpretation for the frequent appearance of the number 10^{40}, though it necessarily draws in the inconceivable aspect of Reality. The Universe—the Whole—is **not** getting any older. Rather, we only conceptualize that it's getting older. Aging—and the "passage" of time—are a profound illusion. In other words, though every limited, bounded thing appears to be getting older, it does not follow that the Universe in Totality is getting older. The Real Universe appears, and can only appear, *Now.* (Einstein once remarked that time is an illusion—albeit a very persistent one.[28]) "Getting older" only involves the relative aspect—the universe as an object. It doesn't apply to the Universe as the Whole. To *see what's* going on, the Absolute aspect must also be accounted for.

And so, 10^{40} is not 10^{40} **because** human beings are here. 10^{40} is 10^{40} because the Universe cannot be, never was, and never will be anything but 10^{40}, *Now.*

So why do we have the sense that things are progressing, that things are changing, that the universe is getting older, that **we're** getting older? How do we account for our apparent passing away? Is it an illusion that we're born, that we live for a while and then we die? Why do we have this sense? We try to account for 10^{40}, but our explanations take on a sort of medieval mentality. If resorting to the anthropic principle does not reach far enough to answer these questions, what will? How do we account for our sense of time if all physical explanation remains inadequate?

ABOUT TIME

M. C. Escher possessed a rare gift for drawing scenes that, when viewed on the small scale, seem quite ordinary, yet, when viewed as a whole, appear utterly impossible. Let's consider his lithograph, *Up and Down*, Figure 7–7. Note the strange quality of the uppermost staircase.

The poor fellows on those stairs may very well forever ascend and descend, but after endless effort they will not have gotten anywhere. This is often how we see ourselves, is it not? We conceive ourselves to be living a rat race where, from day to day, we just go on and on and on, without joy, without wonder, without meaning, without hope. For many of us, particularly for those of us who sense that we live in a fast-paced global village of estrangement, life is no more promising than that of the men in this drawing. Yet, though life may often appear like the scene in this drawing, such a view of Reality is an illusion. This is not *seeing.*

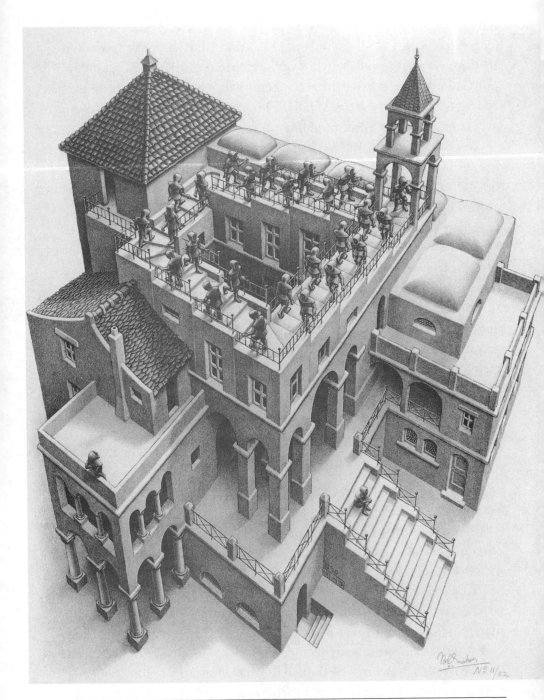

Figure 7–7. Up and Down by M. C. Escher (lithograph, 1960)

At first glance the scene looks quite normal. We see stairs. We see people going up and down the stairs. If we look at one little locality at a time and not at the picture as a whole, everything looks fine. By taking only a small sample, we see order. We can follow the stairs around and find that they keep climbing (or descending). Everything seems normal and understandable—until suddenly we find ourselves right back where we had been before.

This is how we commonly view our lives. At first we do not question. Everything is assumed. Nothing appears strange. Life just is, and we go along. But then, suddenly, we may notice that our life as we have always lived it seems without meaning.

If we look at any one small part, either in this etching or in our own life, everything appears understandable. It is only when we stand back and take it in as a whole that, like a former Flat-Earther looking out to sea, we can see our previous view as an illusion. It is only when we glimpse the Whole that we do not get taken in by the illusion of movement and of time—i.e., by the illusion that our objects Really are things unto themselves.

When I showed Escher's drawing to an eight-year-old, she didn't get it at first. When I pointed out the unusual quality of the stairs, she thought it was "neat." She thought it would be fun to build stairs like that. When I told her that we couldn't build such stairs, she wanted to know why. I explained that the reason Escher got away with this—and why we really can't build such stairs—is that he projected what appears to be a three-dimensional scene onto a two-dimensional surface. He borrowed from an extra dimension, we might say. If such a building were actually constructed—that is, if we were to build in three dimensions a building that could show us Escher's view—such a building could not stand up on its own, for it wouldn't have a back. The **levels** of each stair, when followed around, actually form a spiral. The stairway itself, however, is on a single plane. This becomes more evident when the lithograph is viewed from an angle slightly above the plane of the page. From such an angle you may also notice that the structure seems to rise from the page, which is a common characteristic of such anamorphic figures.

When three-dimensional creatures such as ourselves look at those two-dimensional people living in their two-dimensional world, we can see quite easily that we are looking at an illusion. They, on the other hand, don't seem to fully realize this. They walk up and down the stairs, somewhat long-faced, perhaps even as we would if we had a vague sense that, though time seems to be moving us along, we're going nowhere. They can't see what's wrong because at every little locality (at every small sample) where they "presently" exist, everything makes sense: it is a stairway going up and down. It is when a sense of the Whole comes over them that they must feel the bite of despair, for once they sense their endless repetition upon the stairs, their lives must seem empty and meaningless to them. Yet they do not see enough to stop their useless activity.

When **we** look at them, of course, we realize quite quickly that they're not going anywhere. There's all the appearance of movement and change, yet, like 10^{40} appears to us, everything stays the same when taken as a whole. Everything changes in the local scene (r), yet overall (i), nothing happens.

We in three dimensions have a way of seeing their plight as no more than an illusion. Yet, though we might tell them not to waste their time and energy on such needless activity, we in our three-dimensional world carry on in very much the same manner. We witness the flux and flow of the world, we think things are changing, that we're progressing, that indeed time is moving on. And, therefore, because we don't *see* the Whole picture, we believe that 10^{40} **will** become some other number in the future, for incessant change is all we ever see. We don't *see* that there isn't any future, only an eternal *Now*.

We see things change, and age, and appear and disappear. Trapped in our three-dimensional world, we do not *see* that the fourth dimension, time, doesn't change or go anywhere. *Now* is a constant. Our "aging" universe has only local meaning. Nonlocality, as a Whole, is ageless, appearing always as *Now*. The passage of time is an illusion.

Ten to the fortieth is *Now*. *Now* is all there is. Things appear like they're changing—moving and aging—but we can *see* that this is an illusion. What we do not commonly *see* is the interidentity of past, present, and future.

part III

WHAT MATTERS

Ching Ch'ing asked a monk, "What sound is that outside the gate?" The monk said, "The sound of raindrops."

Ch'ing said, "Sentient beings are inverted. They lose themselves and follow after things."

The monk said, "What about you, Teacher?"

Ch'ing said, "I almost don't lose myself."

The monk said, "What is the meaning of 'I almost don't lose myself'?"

Ch'ing said, "Though it still should be easy to express oneself, to say the whole thing has to be difficult."

eight

⤳(INERTIA)⤳

There are trivial truths and the great truths.
The opposite of a trivial truth is plainly false.
The opposite of a great truth is also true.
—NIELS BOHR

PUTTING SCIENCE TO WORK
The waitress brings me coffee,
I add cream to the cup.
I ask her not to bring a spoon,
Convection stirs it up.
—DALE C. HAGEN

SPIN

As you pour cream into your morning coffee, a beautiful billowy pattern appears. The coffee appears to move slightly in various directions, but on the whole it lies almost still. As you look across its surface, the coffee appears quite flat within the cup.

You take a spoon and stir the cream into the coffee. The cream swirls into the liquid and the coffee turns within the cup. But now its surface is no longer flat. It has become concave instead. As the coffee streams out from the center, being thrown outward by centrifugal force, the center of the coffee becomes depressed. The coffee piles up against the walls of the cup, where the level rises.

How does the coffee "know" that it's supposed to do this? In other words, what "tells" the coffee that it's spinning and should therefore rise at the edges and sink in the center? This seems a ridiculous question at first. But actually it's a very profound question, and one that has kept philosophers and physicists musing for centuries.

Our commonsense response might be, "Well, the coffee 'knows' it's spinning because it's moving past the walls of the cup. It's merely an expression of the relationship that it has with the cup—it's just turning inside the cup, that's all."

This answer seems reasonable enough, but it's wrong. Newton showed us that it was wrong centuries ago. He discovered that if we take a bucket, hang it from a long rope, twist that rope until it's tightly wound, then put water in that bucket and let it go, the bucket will begin to spin. Newton observed that at the beginning the surface of the water is flat, but as the water begins to pick up motion from the bucket through the friction created by the drag along its sides, it too begins to turn. At this point, once again we find that the surface of the water becomes concave. (We can accomplish the same thing today by simply placing the bucket of water, or the cup of coffee, on a turntable and letting it spin.)

But how does the water know when it's spinning and that its surface should therefore become concave? We can no longer claim it's merely the relationship that the water has with the walls of its container, for now the walls of the bucket are turning about with the liquid. **The water is stationary in relation to its container,** yet it still responds as did the coffee when we swirled it in the nonturning cup. It's not merely a relation between the liquid and its container. In relation to what, then, does the fluid spin?

In a nonrotating bucket
the surface of the water lies flat.

In a rotating bucket
the surface of the water is concave.

Figure 8–1. Newton's Bucket

When I turn the coffee in my cup, I cannot conclude that it spins in relation to the cup, or in relation to me, or even in relation to the room. None of these relationships turns out to be correct.

Picture this: let's put a TV camera in a windowless room that is centered upon a large turntable. The camera and all furnishings in the room are securely fastened to the walls and floor. There is a circular pool of water in the very center of the room. We are outside the room—in another building, say—viewing the scene on a TV monitor. As we begin viewing, the room is stationary. The water lies flat within the pool. As the room begins to rotate, everything in the room appears not to change—except the water in the pool. In response to a strange unearthly force, the water begins to depress in the center of the pool, while mysteriously climbing the walls of its basin. In time, as it rises above the sides of its container, the water flows outward from its pool and across the floor. To any viewer watching this scene, only the realization that this room was set to spin could dispel the mystery.

But, from the perspective of the viewer, what authority commands that the water must in one moment lie flat, and in the next moment become concave? We say it's the water's spin. This spin seems to suggest a relationship, but in relation to what does the water actually spin? Not to the pool, or to the room—but what else **can** it spin in relation to? This is the deep mystery that has troubled philosophers and physicists.

THE BOGUS FORCES OF SPIN

The Earth spins, but how do we know that it does? We might point to the fact that the sun, the moon, the planets, and indeed the whole starry sky make complete circuits of the Earth roughly once each day. But why do we say it's the Earth that spins and not the sky that rotates?

It's because there are certain "forces" that become apparent as they act upon the fluids of the planet. There is, for example, the centrifugal force that causes the Earth to bulge outward at the equator. This force, if it were strong enough, would have the Earth and everything on it fly outward to join all that is not the Earth. Things that spin always want to fly apart in this way. But if a spinning object—particularly a large object like a planet—doesn't fly apart, it's because the centrifugal force is predominately countered by centripetal force. In this instance it's gravity that pulls the Earth together and holds it unto itself. The Earth's centrifugal force is no match for the Earth's gravity. In fact, if you were to travel to the equator, where the Earth moves the fastest, the force that would otherwise throw you into space would only be enough to lighten you by a few ounces.

But there's another force, a far more visible one, called the Coriolis force. It's perhaps better called the Coriolis effect, for it's not a real force. The Coriolis effect becomes apparent when we study the large masses of fluids in the Earth's atmosphere and oceans. These large masses of fluids tend to travel in straight lines (as seen by an observer in space), but they are deflected to the right in the Northern Hemisphere and to the left in the Southern by the Earth's spin. And the "force" that causes such deflection, the Coriolis "force," would seem to account for the swirling of cyclones, and the turning of the oceans in their basins. All of this circular motion comes about because it is indeed the Earth that is turning and not the sky that is rotating.

These apparent forces, however, are not real forces. They are merely effects caused by spin. For example, when you're riding in a car that is traveling at a constant speed, if we disregard the effects of gravity (which remain constant), you do not feel any force acting upon your body. If the car accelerates or slows, however, you'll feel an extra force—a changing, momentary pull or push. You'll also feel an extra force acting upon you as the car rounds a curve, even if it maintains a constant speed.

Why do we feel these extra forces when we accelerate or change direction? It's because in both circumstances we experience a change in velocity. Velocity must be distinguished from speed in that velocity is a **vector** (i.e., it's a pointing arrow whose speed **and** direction must be taken into account), whereas speed is merely **scalar** (i.e., direction is unimportant).

We invented this compound unit of measure that combines both speed and direction precisely because it's only when we change our velocity that we feel these "forces" acting upon us. But they are bogus forces. Imagine you're riding in a car once again. To any observer not in your frame of reference (i.e., anyone not moving with you, such as a person standing on the sidewalk and watching your car round a curve), it's obvious that what you feel as a force merely results from your tendency to continue moving at the same speed and in the same direction. (It was through this manner of viewing forces that Einstein came to see gravity as the result of the peculiar geometry of four-dimensional space/time, in which the presence of a mass causes the surrounding space to curve.)

These forces are therefore not merely relative, but in fact fictitious. Yet they seem quite real, for we can readily witness their effects. In fact, these effects can even overtake a "real" force such as gravity, as any toy top can demonstrate. As long as the top is spinning with enough speed, it will not be easily pulled out of line. Gravity is far too weak to take it down. It's this very effect that we use when we ride a bicycle: gravity pulls things down, but a spinning wheel is very difficult to knock over. Even a riderless bike will stand against gravity as long as it's moving and its wheels are turning.

As soon as the spinning stops, however, there goes the so-called force—the still top, the motionless bike lie on their sides, pulled down by the ever-present (real) force of gravity. Only when they spin can they defy gravity. When spinning stops, the bogus forces disappear.

SPIN AND SELF

We still have the question of spin. What is it?

Let's go back to Newton for a moment. Newton, as we saw, demonstrated that it's not the movement of water in relation to the walls of its container that causes the water to respond as it does. As we can see with the Earth, something far more vast and universal than a thing's relations with nearby objects is needed to explain spin. But what? It seems there must be some **Absolute principle** involved.

Newton speculated that things exhibit rotation in relation to what he called "absolute space." But this was not a very satisfying explanation—even Newton didn't like it. Though he held on to the notion of absolute time, he was bothered by the idea of absolute space. After all, we can't tack up a nail in space somewhere to get our bearings from it. What could "absolute space" possibly mean, anyway?

Thirty years after Newton raised the question, however, George Berkeley, noting that absolute space cannot be, in his words, "perceived" (i.e., it cannot be conceived or held as an object of consciousness), stated that it therefore couldn't be used as a reference point either. Berkeley noted that if everything in the universe were annihilated but the Earth, then it would be impossible to imagine any motion of the Earth, including its rotation. In fact, said Berkeley, the notion would be absolutely meaningless.

Berkeley would argue that the Earth spins in relation to the "fixed stars." According to Berkeley, there's an immediate communication between the Earth and the distant stars in the heavens that tells the Earth, in effect, "you are distinct, separate—you are spinning."

We may extrapolate from Berkeley's assertion that if there were no such communication between the Whole and the part, then there would literally be no spin to the large air masses and no turning of oceans in their basins, for without spin these phenomena, these bogus forces, would cease to be. After all, when a fluid body—say, the ocean—turns on the surface of the Earth, it's merely an "attempt" by the moving fluid to continue in a straight line in relation to what remains of the universe. But if nothing remained of the universe beyond the Earth, such relationship would be broken, for such absolute referent would be lost.

Berkeley's ideas in this area seemed quite strange to his contemporaries, who largely ignored them. After all, what could the distant stars have to do with phenomena close at hand? Strange as these ideas sound, however, Einstein, Mach, and others considered, nearly two centuries later, that there might be some merit to them. Since it was Ernst Mach, the nineteenth-century physicist and philosopher, who revived Berkeley's idea, this notion (as it was formulated by physicists and philosophers in the twentieth century) became known as Mach's principle.

According to physicist Clifford Will, Mach's principle asserts that the "inertial and gravitational properties of matter are in some sense linked to the existence of the rest of the matter in the universe."[1] In other words, it's a principle that relates the Whole with the part. (Will cites Newton's bucket as a simple demonstration of this principle.)

Though Mach never properly formulated the principle that bears his name, it was his speculations in this area that inspired Einstein to move along these lines of inquiry. Einstein thought that proof of such a relationship between the Whole and the part was buried within his general theory of relativity (though late in life he gave up on the idea because of the mathematical enormity of the problem). According to developments in mathematics in the 1980s, however, it appears that there might be some merit to this idea.

While the idea of absolute space and time has been abandoned by science in modern times, the idea of absolute motion has not. Absolute motion, however, simply refers to acceleration and rotation. But these "absolutes" are not really Absolute in themselves. They merely express the relationship of Whole with part.

The quality of an object's spin, we now know, has everything to do with its inertia. Inertia is defined as an inherent property of matter in which, according to Newton, "a body continues in a state of rest or constant velocity unless acted on by an external force." An object is said to automatically oppose a change of motion by reason of its inertia.

Inertia is a relationship that a physical entity has with the Whole. This relationship "tells" an object how far it should slide when pushed. That is, a thing "knows" how far to slide because of its relationship with the rest of the universe—including the stars in the most distant galaxy. (We no longer refer to them as "fixed stars," of course, for we now know that stars are grouped into swirling galaxies, and that all matter in the universe is in relative motion.)

We might say, then, that spin is the delineating relationship between a thing or idea—i.e., any conceptualized entity (*r*)—and everything else that exists (*i*). It's what gives unity to things and sets them off from others. Spin is what distinguishes the part from the Whole, for it can't be maintained that the Universe as a Whole spins. Wholeness, Totality, can't spin; without an external referent, such a quality is without meaning. **Spin, therefore, is the essence of the relationship between the Whole and the part.** In short, it's conceptualization.[2]

But, just as a viewer outside a swerving car sees no "force," but only the tendency of the car and its occupants to travel in a straight line, so too when "viewed" from the Whole, where we *just see,* where there is no spinning, no self appears. There is, in fact, no separate entity of any sort that sits apart from others, or from the Whole. Like the bogus forces of spin, a self appears in conscious awareness only as something that has been defined as separate and distinct from other—"other" being, finally, the rest of the universe.

Roger Penrose noted in *The Emperor's New Mind* that the amount of angular momentum found in quantum objects indicates that such objects cannot be "composed solely of a number of orbiting particles, none of which was itself spinning"; rather, such momentum

> can only arise because the spin is an **intrinsic** property of the particle itself (i.e., not arising from the orbital motion of its "parts" about some centre).[3]

In other words, spin is not, in the final analysis, a phenomenon that can be explained in terms of bodies orbiting about some center. Rather, physicality (indeed, all conceptual reality) is the **result** of spin—the relationship of Whole and part—itself. In other words, spin—i.e., that which marks the relationship of the part with the Whole—is more fundamental than physicality itself.

THE GRAND SYMBIOSIS AND OPPOSITION

Just as the spinning Earth cannot exist by itself as a **spinning** Earth, so too a self cannot exist on its own, without a dynamical relationship to a larger whole. There's a grand symbiosis at work in Reality. It's this grand dynamic that occurs between the Whole and each part. And it's a relationship that we can demonstrate physically, as in the case of a spinning top, or, on a grander scale, a graded stream.

A self cannot persist as a static entity, an absolute. It can only appear as something akin to a spinning dynamo—as something we cannot get our hands on, so to speak. At the same time, however, when something spins, it's rendered distinct and separate from everything else. In other words, spin, or conceptualization, **is** the manifestation of things as they appear in opposition to all other.

When I was young, it was required of all boys to take a semester of printing in the eighth grade. It was there that I learned where the expression "watch your p's and q's" came from. It was a warning our teacher drummed into us because, as we removed the type from the galley to put it back into the case, all of the letters appeared as their mirror images. So when we encountered a "p," it looked like a "q," and the "q" looked like a "p." Therefore, when we put them back into the case, we had to be careful not to mix them up. We had to watch our p's and q's.

With that image in mind, let me present a scheme of mirror opposition:

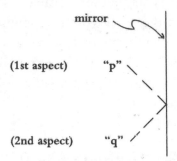

Figure 8–2

The vertical line represents a mirror seen on edge, and the letter "p," being held before the mirror, projects the letter "q." The letter in the upper position ("p") I will refer to simply as the first aspect of mirror opposition. This position can represent the subjective—"me," "my view," "our view," etc. What appears in the lower position ("q") I will refer to as the second aspect of mirror opposition. This position represents the objective—it's what the "I," the "we," or a self sees as being "out there." This scheme of mirror opposition is meant to represent our common-sense view of the world, not the way the world actually is.

When conscious awareness is dominated by common sense (i.e., by conceptions of self and other), we see mind-objects (things and ideas) as though viewing them in a mirror. At the same time, however, we do not recognize the mirror but, rather, assume we're looking through a window out onto a world "out there." We are not aware that we're merely conceiving—and, in fact, projecting—that thing which appears to be "out there."

When conscious awareness is caught up in the sights and sounds of the phenomenal world (particularly the material world), we're likely to assume that when a "thing" appears "here"—such as "me" (first aspect)—then whatever appears "there," in the mirror, is seen in opposition to "me." In other words, it's seen as being quite different from "me"; since it's seen as being "over there" (second aspect), it's never seen as being a reflection of "me." Rather, "that thing, out there" is seen to be what **opposes** "me."

Thus we commonly conceptualize the world. That we do so is exemplified in the conflicts that arise between you and me, male and female, good and evil, etc. These opposing conceptions are basic to our usual ways of interacting with one another.

This opposition arises both in situations of antagonism and in situations where we desire our object. In these latter cases, we first conceptualize the desired objects as existing separately "out there." Then we try to pull what appears as "apart from me" into a position where it might appear closer to "here." We attempt to identify with our object—a most frustrating task.

Whether, however, the situation is one of antagonism or desire, we're not seeing things as they are; we're missing something. We're not *seeing what's* going on. We're not *seeing* the Whole.

We're missing what we might call the third aspect, which doesn't reside in either the concept of self (me, here) or other (that, over there). **It lies outside these two alternatives.** Mirror opposition always implies this third aspect of Reality, which common sense habitually overlooks.

Using the same schematic, this third aspect can be shown like this:

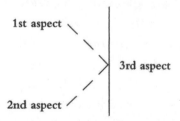

Figure 8–3

This third aspect in fact opposes **both** the first and the second aspects taken together. I shall call this subtler relationship ultimate opposition.

A very simple metaphor can be used to demonstrate this sort of opposition: our Earth. If we were to look down upon the Earth from a point in space directly over the North Pole (our view being the first aspect), the

Earth would appear to rotate counter-clockwise. Its mirror opposite (the view from over the South Pole) would have the Earth appearing to rotate clockwise. Here we have clearly opposite views. But these two opposites, taken together, comprise the spinning Earth itself: a single, unified entity. This spinning planet exists in opposition to any single simplified view of what direction it turns. The Earth itself reveals a third aspect—i.e., ultimately it opposes either view of clockwise or counter-clockwise.

A mirror scheme showing three aspects of the Earth as a spinning system can be diagrammed like this:

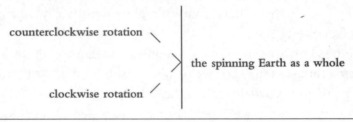

Figure 8–4

Thus we can see that mirror opposites are actually dual aspects of a single entity—in this case, the Earth. If we discuss the Earth as a rotating system, we must see the mirror opposition of the poles, but we must also realize that these two poles imply each other. In doing so, we must also see that they imply a third, "ultimate opposite": the Earth itself, which draws these opposing mirror images into a unit.

All of this is clear enough in the example of the Earth, or some other physical entity; but, as we shall see, such metaphors are limited. What I actually mean by "ultimate opposition" is that the third aspect of opposition is Absolute. It's inconceivable—i.e., it doesn't form as an actual object of mind. Nevertheless, ultimate opposites—that is, the pattern of ultimate opposition—can always be found, if only we would *just look* beyond our commonsense view of mirror or conceptual opposition.

What, then, is the third aspect that is implied by a moral mirror opposition—say, good (aspect 1) and evil (aspect 2)? We don't have a term for it. To ordinary human conscious awareness, to common sense, there **is** no third possibility, for good and evil are not commonly seen as a single unitary system. Yet such a third aspect can be *seen* if we *know* how to *look*.

ULTIMATE OPPOSITION

Ultimate opposition most vividly contrasts with mirror opposition in that it causes harmony rather than conflict to arise. Ultimate opposition is, in a word, generative.

To illustrate this, let's consider the following example: if I draw a line like this:

$$\underline{\quad\rightarrow\quad} \text{ (from left to right)}$$

and ask while I draw it, "What's the opposite of this?", you might understandably think:

$$\underline{\quad\leftarrow\quad} \text{ (a line drawn from right to left)}$$

But this is simply the same line drawn as its mirror opposite. We have basically the same thing in either case. With mirror opposites we always end up with the two being essentially the same, for what we have is merely a single **thing**, ———, (i.e., an object that makes use of the *r* aspect alone) reflecting itself, rather than a picture in which the "remainder" of Reality (i.e., *r* + *i*) is accounted for. Conceptual (i.e., mirror) opposites, therefore, are not ultimate opposites.

Ultimate opposites account for both the *r* and the *i* aspects of Reality—and these two aspects of opposition do not resemble each other in a mirror. In other words, the contrast is not between +1 and −1 (as we might find in mirror opposites), but between *r* and *i*.

Our common sense usually places opposites as though they were in some sort of confrontation, like an arrow reflecting itself in a mirror:

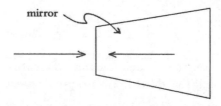

Figure 8–5

Ultimate opposition, however, generates peace and harmony instead of conflict.

If we are to model ultimate opposition, we need a nonreflecting image. I therefore suggest that we depict the ultimate opposite of ——, the steady, unchanging line pictured above, as 〰, a line of change, curve, and oscillation.

As we've learned from Schrödinger, Lao Tzu, and Thích Nhât Hanh, by our very conceiving of an object as a separate thing, that mental object immediately begins to spin with its conceptual opposite, like the infinite swirls between the "inside" and the "outside" of the Mandelbrot set. These two concepts come together and spin about each other in an exchange of identity. We can illustrate this in our mirror scheme like this:

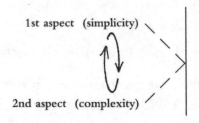

Figure 8–6

When something spins it has an axis, and its axis suggests a direction that points **beyond** the spinning system, much as north points beyond the Earth, while east and west do not. This is the quality of spin that indicates or implies a third aspect:

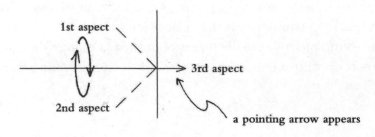

Figure 8–7. The spin, i.e., the mutual exchange of identity of aspects 1 and 2, forms an axis that points to a third, nonconceptual aspect.

Our spinning objects of consciousness thus point to something other than (and ultimately opposite to) themselves.

Thus we see a pattern that characterizes the two elements of our experience of Reality, the relative and Absolute. The relative aspect—conceptual reality—is characterized by spin. If we latch on to any object within that world of spin, we too begin to oscillate, back and forth, up and down, with the spin. The other element of experience, however, is steady and unchanging. This is the aspect of experience that is Absolute. This pattern—the pattern of ultimate opposition—must be *seen* in any True account of experience.

We see this pattern of ultimate opposition in how we use language. In order to communicate verbally, two elements or principles must be involved. One is steady and predictable, while the other is novel and unexpected. Either function by itself will not do the trick. If what is being transmitted were totally unexpected, there could be no communication, but only gibberish. On the other hand, if what is being said were totally expected, again there would be no communication, but the expression of what is already understood.

Jeremy Campbell, in *Grammatical Man*, wrote that

> a written message is never completely unpredictable. If it were, it would be nonsense. Indeed, it would be noise. To be understandable, to convey meaning, it must conform to rules of spelling, structure, and sense, and these rules, known in advance as information shared between the writer and the reader, reduce uncertainty. They make the message partly predictable....[4]

On the other hand, says Campbell,

> a message conveys no information unless some prior uncertainty exists in the mind of the receiver about what the message will contain. And the greater the uncertainty, the larger the amount of information conveyed when that uncertainty is resolved.[5]

Thus we find that language lives in this dynamic, in this pattern of ultimate opposition.

We also find this same generative and cooperative pattern of ultimate opposition in the way sex chromosomes combine. While our common-sense view of opposition might have us think that if the female carries a p, then the male must carry a q, this is not the pattern we actually find. Females, as we know, carry two X chromosomes, while males carry an X and a Y. Thus, while the female provides the steady element (she can only contribute an X to her offspring), the male oscillates by providing either an X or a Y to his progeny. Thus, again, we find the pattern of ultimate opposition functioning in the way the female and the male come together to generate life.

We'll be looking for more signs of this pattern in what remains of this book, but first we must take note of the danger we face when we try to conceptualize the inconceivable.

THE HIERARCHY OF OPPOSITES

Usually, when the mind attempts to grasp the first two aspects of opposition at once, rather than just watch them spin, the mind itself begins to spin. The mind is then characterized by conflict. To this mind, the two aspects can (and usually do) go after each other and turn about each other in competition or conflict—often leading to paradox and confusion. We don't realize that all of this competition and conflict can be done away with in an instant of *seeing*—for, when the third aspect is drawn into view, what had been characterized by conflict suddenly appears as harmony.

This harmony, while impressive—often impressive enough to make the person who experiences it believe they have experienced enlightenment—is typically only temporary. This lack of permanence has to do with the way the mind works. As we have seen, the mind creates a bogus ego consciousness, and this consciousness is very quick to make its objects its own. The result is that the newly found third aspect is often adopted or identified with by this consciousness. When this happens (and it usually happens immediately), what had been *seen* as the third aspect becomes conceptualized by this bogus consciousness as a new first aspect.

The latching onto the third aspect and pulling it into the first position usually occurs without our recognizing what has happened. In other words, once we suddenly realize that our interests are inextricably linked to the other guy's, and we realize that this is a much deeper or more profound way of seeing things, we think, "This is it! I've got it!" At this point, we've just pulled our insight down into a first aspect position. Yet we fail to notice that we're still clutching a static view. We've identified with our discovery and turned it into the first aspect, and it will only be a matter of time before this view will reflect yet another view, and another second aspect will emerge. Conflict will then, once again, arise.

Each new awareness not only strikes us as being quite real, but as being **more** real than our previous understanding—much like we might sense that the planet Earth is of a greater or "more real" nature than either of its poles. Yet, just as this does not mean that the poles are **not** real, so too the collapse of the first two aspects does not mean that what was previously conceived as real in some way disappears. We still see the tapered tiers of the café-wall illusion even after we know their defining lines to be parallel, but we now accept our earlier impression as a naïve, less tangible, or less valid view. Rather, what occurs at the "collapse" is that the third and heretofore unnoticed aspect of *what's* going on makes an indelible impression. Such change always strikes with utter clarity and apparent certainty, and does not appear uncanny in the least. Unfortunately, we're likely to take this to be *knowing*, the direct *seeing* of a higher order of Reality.

But True Knowledge is not in grasping yet another object, but in *seeing* the hierarchical structuring of mind—that is, *seeing* what the mind does and how we package our perceptions.

Thus we can begin to see a hierarchy of opposites forming in the mind. There is no end to this chain of mirror opposition, and it's very deceptive, in that at every step one is made to feel that they are growing spiritually or, at least, intellectually. But moving up this chain is really an illusion, for it actually leads nowhere.

It's crucial to *see* this process—not because there is any hope of ever arriving at its end, but in order to step off this endless chain of apparent hierarchical levels that resembles the illusion we found in Escher's stairs (page 212). It's in *seeing* this process (*seeing* the pattern) that we can get a glimpse of ultimate opposition, of *what* is Really going on.

SCRUTINIZING THE THIRD ASPECT OF OPPOSITION

Most of the time we remain oblivious to the third aspect, which hovers just above the plane of whatever conceptual level we're on—except for that brief flash of insight that hits from time to time.

So let's look at our opposition scheme again and plug in a few common frustrating experiences, and see if we can pull the third aspect into view, so that we may better *see* the pattern.

Let's return to our problem of good and evil, and make use of the common white hat/black hat analogy that originated in old movie Westerns. In these movies, the good guy wore a white hat, whereas his mirror opposite, the bad guy, wore a black one. We have here one of the simplest, most clear-cut mirror oppositions, graphically illustrated:

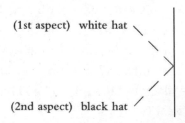

(1st aspect)   white hat

(2nd aspect)   black hat

Figure 8–8

But what is the True Opposite (third aspect) of both of these fellows? It's the man who, metaphorically, wears no hat at all.

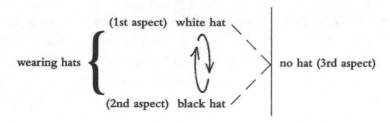

Figure 8–9

The person who wears no hat is the person who's not taking sides—the person who does not see themselves in opposition to others.

The black hat/white hat view of good and evil can make an entertaining movie because the moral lines it draws are so simple and obvious that the story remains easy to follow until its poignant finish. Outside of movies, however, the lines are infinitely complex and the story has no ending. The black hat/white hat theory of good and evil doesn't reflect our actual experience of life's moral difficulties.

The common view is to think: **we** are good, and evil is out there. We label those other guys out there as evil because we're already assured, through our tacit assumptions, of our own goodness. Yet we unwittingly do what they do. They build their weapons to defend themselves against us, and we build ours for the same reason. They claim God is on their side and so do we. Our everyday sense of good and evil thus only helps to generate conflict.

As long as we behave like this, it's clear that we still do not understand the nature—the sameness—of what we are calling good and evil. **Good—True Good—does not resemble evil in a mirror.**

True Good, as we can sense in our hearts, must exist in a totally different way—like this:

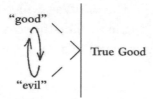

Figure 8–10

True Good—those actions that spring from Awareness of the Whole, from *just seeing*—is utterly beyond any everyday sense of good and evil, or right and wrong, or pleasant and unpleasant. It's the appropriate action at the appropriate place and time, free of selfishness, attachment, conceptualizing, and even (as we normally think of it) intent.

———————

Let's consider how we commonly deal with anger. In our scheme of simple mirror opposition, dealing with our anger normally looks like this:

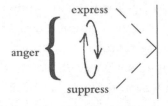

Figure 8–11

"Express" in the first aspect means we hold the view that it is best to express our anger. The mirror opposite of "express" is "suppress," which is the second aspect.

Years ago, we used to think it wasn't good to express our anger. Since then, we've come to learn how unhealthy it is to suppress it. If we suppress our anger, or deny it, we become neurotic. Worse, we might eventually blow up like an over-heated pressure cooker.

We're now more inclined to let our anger out. We might go to a therapist and beat a few pillows to let off steam. It feels good, of course—for a while. But ultimately it's equally unhealthy, because our anger doesn't really diminish when we do this. Worse, we're likely to start expressing our anger on other people. When we do, of course, they're just as likely to blow it right back in our face, which only serves to make us even angrier. Soon both people develop the sense that their anger is justified. This anger then grows and becomes ever more real, ever more "justifiable."

What we don't usually understand is that there's a third aspect. (Indeed, there's **always** a third aspect when we're dealing with the world in concepts.) This is to *just see* anger and identify the pattern being generated by such conceptual opposition. *Just see* the anger. Don't feed it, don't justify it, don't do anything to it. Don't try to squelch, limit, or deny it, either. *Just see* it for what it is.

To *just see* anger, and to discover the pattern, is to avoid pulling anger down into the first aspect.

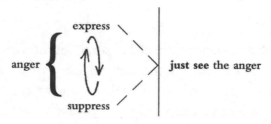

express

anger { } just see the anger

suppress

Figure 8–12

When you do feel anger, as quickly as possible just go to the third aspect—recall it—and then just observe the anger. If you choose to express it, *see* what happens when you express anger. If you stifle it, again *just watch* what happens. Just keep observing the anger itself and notice *what's* going on. Don't be too concerned about whether you're suppressing or expressing. *Just watch* and *see* what follows what.

Just seeing is, in fact, a most effective way of dealing with any problem that grips us emotionally.

What else can we view with our opposition scheme? Anything we can conceive:

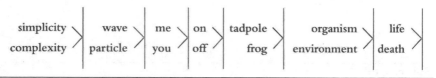

simplicity	wave	me	on	tadpole	organism	life
complexity	particle	you	off	frog	environment	death

Figure 8–13

Let's look at one more example. Progress is an idea that lathers up a lot of people these days. How would our commonsense view of progress appear in our three-aspect scheme of mirror opposition?

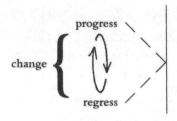

Figure 8–14

When progress is seen as opposed to some sort of regress, or a going backward, just what exactly is it that is going forward? We've already discussed several instances where our science and technology, the hallmarks of our modern age, appear in many ways to be little more than open warfare on the ecology of the planet. I will not enumerate more examples here, for they can be found in abundance elsewhere. The point I want to make is that what we commonly think of as progress may sometimes be counter to what we commonly mean when we use the term.[6]

What would be the ultimate opposite—the third aspect—of change? Instead of attaching to our ideas of moving forward or backward, it would mean *just seeing* change for what it is. This reveals nothing resembling progress (or regress) at all. In fact, as we have already seen in various ways, the *i* aspect of Reality does not reveal change at all.

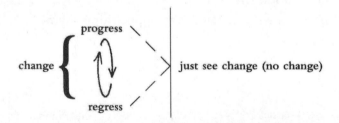

Figure 8–15

When we put ourselves in conflict with things and ideas—when we believe we must get "out there" and fix things up—it means that we're not *seeing* in whose hands such powers lie. When we believe we are acting for the good through our good intentions (i.e., our concepts of goodness), it means we do not *see* that we are only drilling holes into the face of King Chaos.

It might seem at first that to *just see* is impossible—or that doing so makes us stupid, or dull, or naïve. But this is not at all the case. In fact, *just seeing* is the highest expression of a rational mind. It marks the truly open, fluid, tolerant, magnanimous mind that is Wisdom.

The mind that subscribes to no view is a ready mind. It's ready to receive whatever comes before it. It's necessarily a mind in full Knowledge of ignorance. The mind that is aware of what it doesn't know is not the common, fearful, rigid mind that desperately seeks to neatly fit the world into its fond and familiar thoughts. Such a mind is open to unadulterated Truth.

nine

⤜(BECOMING)⤝

Let go of the idea, "I exist."

—IKKYU

*Just as a man shudders with horror when he
thinks he has trodden on a serpent, but laughs
when he stoops and sees that it is only a rope, so
I discovered one day that what I was calling "I"
is not apparent, and all fear and anxiety vanished with my mistake.*

—GAUTAMA

NOTHING FROM SOMETHING

What can we notice about the interplay between *nothing* and *something*—between what we might think of as "being" and its inevitable reflection, "nonbeing"?

First of all, we'll find that they're unstable. As soon as one appears, it reveals its other. *Something* reveals *nothing,* and *nothing* reveals *something,* immediately.

Bertrand Russell wondered, "Why is there something rather than nothing?" He was certainly not alone in asking this question. But *something* and *nothing* don't really exist in the way Russell puts it—that is, as →←, or mirror opposites. They don't really look like each other in a mirror, as we shall see. The fact is, they're ultimate opposites—and, as with all opposition, they must both occur at once. Getting through this ultimate paradox will be the thrust of this chapter.

What is it about "something"—the many interrelated and interdependent "things" that comprise our universe—that in fact differentiates it from *nothing,* and that does so in a way other than mere mirror opposition? In other words, what makes *something* ultimately opposed to *nothing?* "Something"—the multiplicity of objects—belongs to the relative realm, the world of thoughts and things. In fact, it **is** that world. The basic characteristic of all entities in this world is that each exists in relationship with all others. We might also call this realm the "world of difference," because relationships invariably reveal differences (if there exists no difference, there also exists no separateness, and thus no relationship).

What lies outside of the world of difference, outside the realm of thoughts and things? *Nothing.* Absolutely *nothing.* In other words, while "something" is always relative, *nothing,* when it dances with *something,* indicates Absolute. It therefore should not be confused with any idea we may have of nothing—such as when we open the cookie jar and find it empty.

Earlier, you'll recall, we noted that Einstein mathematically demonstrated that the universe is finite, yet boundless. Not even space, not even time lies outside the Universe. There's no boundary that divides a "something" here from a "something else" over there. There's *nothing* over there. In fact, as we have *seen,* there **is** no "over there."

The apparent paradox, then, is that the realm of unity **is** the world of difference. But this paradox appears only in the relative, conceptual world. It doesn't appear to direct perception.

SOMETHING FROM NOTHING

Let's take another stab at this notion of *something* versus *nothing*, but this time from a different angle.

I've mentioned that Reality has two aspects, which I've referred to as "here it is" (*r*), and "what is it?" (*i*)—or as *this* and *what*. Or, to put it more simply, *this* and pure ?—pure interrogative. If we pick something up, say "*this* cup," we tacitly assume, "here it is." But now that we've picked it up, if we are perceptive, we'll notice that we've also picked up the question, "what is it?" You'll recall from Chapter 3 how Thích Nhất Hanh showed us how the remainder of the universe can be found in the identity of "*this* cup." Indeed, if we begin to analyze a cup in an attempt to find what it is, it will soon give up its "cupness"; in fact, under close scrutiny, **any** object gives up its "objectness," its separate identity, its self. Finally, what's left? *What* is left. In other words, *nothing* is revealed.

On the other hand, if we take this very *nothing* and examine it very carefully, we find the conceptual world. Indeed, according to the laws of physics, the "something" that is our universe literally came out of *nothing*. And the *nothing* out of which our entire universe came is very much like what scientists call a "quantum vacuum."[1]

A quantum vacuum is not an ordinary vacuum. It's more like a frothing, surging sea of virtual existence. At any given moment, countless absurdly small particles are popping into a somewhat ghost-like existence; then, immediately—within billionths of a second—they pop out of existence again. This is what scientists agree is going on "outside" of space and time. At any moment, however, due to the Heisenberg uncertainty principle, one of these little virtual particles might be caught "outside" the surging froth (that is, it might exist "within" the universe) for too long, in which case the surging froth will then close behind it, leaving the particle stranded in existence (kind of like the fate of the little lame boy in the story of the Pied Piper). But upon being caught in existence—that is, conceptual reality—a physical entity (or any conceptual entity, really) immediately draws the entire universe into relationship with it. In other words, it begins to share (or define) its identity with all it is not.[2]

Thus, as we conceive an object, whether it be purely mental, or Thích Nhất Hanh's cup, or an elementary particle, we also find the "birth" of the universe erupting from what is otherwise *nothing*.

It's in the very conceiving of an object by consciousness, in the very birthing of conceptual reality, that contradiction and opposition and paradox occur. For the creation of an object also creates the immediate entanglement of that object's identity with all that it is not.

MUCH ADO ABOUT SOMETHING

What is, is "something." "Nothing" is what is not. Let's look at these two related commonsense notions.

To do this, we'll take "something" and plug it into our opposition scheme by placing it in the first aspect—thus illustrating our common-sense view of things:

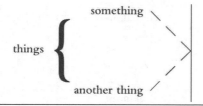

Figure 9–1

In taking this view, we generally believe without question that *this* (our object) "is what is," and that "this is" is Reality. We may or may not notice the other "something"—here called "another thing"—which appears as the reflection of the "something" we presume is "here."

As long as we remain locked into conceptual thought alone, this "another thing" can take a variety of forms. If our "something" is the coffee cup on the table, "another thing" could be the coffee in the cup, or the papers being held in place under the cup, or the table upon which the cup sits, or the room that surrounds the cup, etc. All of these can be mirror opposites of our cup, depending on the context in which the cup (as a concept) appears.[3]

The view that appears in Figure 9–1—that is, our most basic and un-spoken assumption of Reality (that the things we conceive exist in an

Absolute way)—we'll call "eternalism." This is immediately and tacitly what we assume about our experience. Like Huai-jang (page 13), we assume that our objects actually **are** something.

Most of us simply stop all consideration of the question of something vs. nothing with this tacit assumption of eternalism. We don't even consider the possibility that there could be any other description of Reality. In our everyday life, when we lift the cup to take a drink, from our commonsense perspective we've already assumed without even so much as a **first** (let alone a second) thought that the cup "just is."

If, however, we question the possibility that "things" might not be at all—and, remarkably, there are those who **do** believe this—then we've created the mirror opposite of eternalism: nihilism. In our mirror scheme it looks like this:

Eternalism is simply our commonsense view that "something is." It quickly and easily reflects the opposing view: that "nothing is."

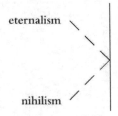

eternalism

nihilism

Figure 9–2

But in the relative world in which we live, all entities—thoughts and things—are always found to include one another, to snap back on each other. They're always found within a matrix of Totality, spinning about one another, and invariably creating paradox.

As an example, here's the heterological paradox first devised by the German mathematician Kurt Grelling in 1908. The word "heterological" was invented to describe words not true of themselves—e.g., the word "long," in being a short word, is heterological, for it's not true of itself. To provide all the necessary terms for the creation of a paradox, the word "autological" was also invented to denote words that are true of themselves—e.g., the word "short" **is** autological, since it **is** true of itself.

So here's Grelling's paradox: Is the word "heterological" heterological or autological? If the word "heterological" were true of itself, it would be autological, but in being autological it would **not** be true of itself, hence it must be heterological, which, of course, means it **is** true of itself, etc., etc., etc.

But now consider this question: Is the word "autological" heterological or autological? Notice that this question does not set the mind to spin, for the word "autological" is indeed true of itself.

We've seen this pattern before. One element, the word "heterological," oscillates when turned on itself; the other, the word "autological," does not.[4] This is the same pattern that appears with ultimate opposition.

And so it is between *something* and *nothing*—which we must not confuse with our **ideas** of something and nothing. It's quite apparent that—like any conceptual pair—one implies the other (hence the inspiration for Russell's question). But these are merely **ideas**—concepts. Naturally, they spin. But we're not talking about our **ideas** of Reality here. We're concerned with direct perception. If we just crawl down into the bare bones of perception, we might notice that one—i.e., "something"—spins, while the other, so long as we don't force a handle on it, does not. In fact, we might notice that "something" spins all by itself, for its "other," even if we think of its other as "nothing," is conceived as just another thing. But this "nothing" is merely our **idea** of nothing. It's like the nothing inside the empty cookie jar, which contains air, crumbs, dust mites, etc. It's not really *nothing*.

But why attempt to grasp either opposite and hold it close, insisting that this—or, indeed, any—scheme explains what's directly perceived? We can instead *just look*. If we just focus in on the bare bones of experience, we can *see* that, indeed, **"something"** is going on. Even when we **do** seem to find a "nothing," such as when we open the cookie jar and find it empty, we still experience something—not cookies, but another thing. There's air in the cookie jar. Even in the vacuum of outer space, there's not *nothing*—there's still space. We do not, we literally cannot, experience *nothing* as a mind object.

What I wish to explore now is the way we actually experience the *something* that is always "there" in experience—indeed, that is ever there

as experience. This *something* is not merely "there." Reality is more ineffable than that.

The Zen monk Huai-jang finally realized that "even to say it is something doesn't hit the mark." But, obviously, saying it's nothing doesn't hit the mark, either. Even saying "there's a lurking something behind the phaneron" (see page 50) doesn't hit the mark. So what in the world **can** we say?

It's clear that we do not—we cannot—experience Absolute directly, but now we must take note that we do not experience the relative directly, either. Neither of these aspects of Reality appears isolated and on its own; rather, they occur at once, together, in the same time and space, the same moments and events. We experience each through the other's filter, so to speak.

Therefore, contrary to Russell's widely held, commonsense assumption, I suggest that we do **not** experience "something" **as opposed to** "nothing." We experience *something* and *nothing* together. We experience (perceive) Reality directly through the **dual** aspects of *something/nothing—r + i*—at once. That is, we experience something as the complement of *nothing* and vice versa. The *nothing* aspect of experience is immediately "there" along with *something*. Neither aspect is more real than the other, but both are necessary for perception. If we attend to just the bare bones of direct experience, we can *see* that this is so. It is much like seeing—**and regarding it as nontrivial**—that a cup is as much implied by what it isn't as by what it is.

Thus *something* and *nothing*—mirror opposites when imagined as two, come together—indeed, must **be** together in a single identity. And thus are we liberated from endless questioning: Is it possible that this cup doesn't exist? Or does it exist, but in a way we cannot account for? All such questions now become meaningless.

THE MARKS OF EXISTENCE

What does it mean to exist? When we say that something exists, we assume and imply that it lasts—i.e., it continues from moment to moment; it persists. According to our commonsense view, a thing is what it is ($A \equiv A$). If it's A at noon and it's still A at 3 p.m., it existed (persisted) for three hours. If it didn't persist, it would no longer be A.

Yet what "something" in fact persists? Everything changes. Until its complete disintegration, a house needs constant repair. Thoughts come and go. Feelings well up and subside.

So, how can something change and yet remain what it is? How can "it" become something else and still be itself at the same time? In fact, what does "it" even refer to?

If things persist (i.e., exist) as themselves, then they can't change. And if they change, then in what sense are they still themselves?

Here's a common response to these questions: "Things persist **and** change. I was once a child, but now I'm grown. I changed. In fact I'm changing right now, yet I persist. That's existence."

But in what sense does the word "I" (or "you") refer to that child? Everything about that child no longer exists. If "you" refers to the body, all the physical elements of that child's body have long ago dispersed into the environment. If "you" refers to the mind, every thought, feeling, or mental impression belonging to that child has long ago vanished.

Sure, you have memories—and photos, videos, recordings, etc.—of that child. But the child no longer persists. There's just the immediate "you," who is obviously not the same as the vanished child.

If you examine this carefully, you'll realize that the immediate "you" is also not the same as "you" of one year ago, or one week ago, or even one moment ago. "You" never persisted; this "you" keeps continually changing.

In fact, there **never** was a persisting "something" to which the word "you" could apply. Nor is there one *Now*.

In other words, "you" doesn't refer to anything that can be pinned down as itself. "You," "me," "I"—these terms do not refer to any existing entity, but only to thoroughgoing change itself.

Put simply, we don't actually **have** things in any objective, substantial way. We just **believe** we do. It's a useful fiction, to be sure, but it's a fiction nonetheless.

Of course, existence is difficult to doubt. In fact our common belief in existence is all but indelible. Even Descartes found it impossible to doubt. To doubt existence would indeed be Great Doubt. Nevertheless, this is what we need to do.

To get beyond this impasse, then, it might help to ask this: If things and ideas exist as we commonly suppose they do, then what marks existence? What are the traits or signs by which we can identify existence?

The first and most obvious characteristic of the elusive **idea** of existence is incessant change. Nothing "exists" without change—not even space or time.[5] Indeed, science has shown that as soon as space itself is defined (i.e., contains something), it appears to expand—to change. Change is the very character of both time and space.

Nowhere in the universe are things or thoughts found to "exist" in an unchanging state. Nowhere is there anything with an abiding individual identity, or its own being. Thus change, the first mark of existence, reveals the inseparability of *something* and *nothing*.

We can find no exception to this rule. All things come and go, for all things come conjoined—indeed, they are interidentical—with what they are not. There exists no solid, permanent, stable thing at any point. There's only a constant stream of interconnected, interdependent, or conditioned events. In other words, change doesn't mean that a "thing" becomes "another thing." That's only the **appearance** of change. Change is the total dynamic working of interdependence.

In fact, if we attend carefully to immediate experience, we will *see* that, in addition to the fact that *nothing* is ever found to persist, *nothing* is ever found to satisfy, *nothing* is ever found to possess a self.

These three marks of existence—impermanence, insatiability, and insubstantiality—are inextricably linked. We want to endure, but find only change—two characteristics of "existence" that inevitably join with dissatisfaction.

Change is not satisfying when we believe in existence because we inevitably strive to maintain what we appreciate or desire, and to avoid what we don't appreciate or desire. But we cannot reliably or consistently do either, for everything is always changing. We cannot escape change.

Our usual way of dealing with this underlying dissatisfying aspect of Reality—i.e., attempting to maintain what we want and to avoid what we don't want—is doomed to failure. Dissatisfaction is intimately linked to change.

The other mark of existence that joins with dissatisfaction is found in clinging to what is utterly conditional—what we conceive of as a self.

A SELF CANNOT BE APPREHENDED

No matter how hard we try, we cannot find anything that we can refer to as "myself." With everything changing all the time, nothing permanent can be found within our relative world—including our own bodies, our own minds, and our own "I"s. We imagine (that is, we conceptualize) that a permanent, abiding self exists. But no such abiding thing actually appears in our experience.

I would like to return to Descartes' *cogito* for a moment. Descartes clearly did not grasp the fact that he had already assumed a self even before he spoke. The "I" had already been assumed even before he said, "I am."

Starting as Descartes did, and indeed as we all must—in ignorance— what is our first, most direct experience? Our commonsense impression of Reality assumes that there's something that corresponds to the word "self." *"Sum ergo cogito."* I exist, and therefore I think. And then we tacitly assume an external world as well:

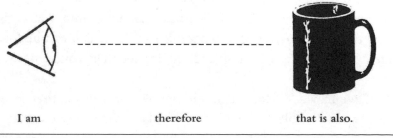

I am   therefore   that is also.

Figure 9–3

But this is **not** a report of actual, direct experience. We do not **experience** an I—we assume it. We only **experience** perception, thought, and consciousness. Just as there's no lurking "something" behind the phaneron, there's no thinker (no self, no "I") to be found behind the thought.

In *The Emperor's New Mind*, Roger Penrose asks,

> What is it that gives a particular person his individuality? Is it, to some extent, the very atoms that compose his body? Is his identity dependent upon the particular choice of electrons, protons, and other particles that compose those atoms? There are at least two reasons why this cannot be so. In the first place, there is a continual turnover in the material of any living person's body....
>
> The second reason comes from quantum physics—and by a strange irony is, strictly speaking, in contradiction with the first! According to quantum mechanics...any two electrons must necessarily be completely identical, and the same holds for any two protons and for any two particles whatever, of any one particular kind. This is not merely to say that there is no way of telling the particles apart: the statement is considerably stronger than that. If an electron in a person's brain were to be exchanged with an electron in a brick, then the state of the system would be **exactly the same state** as it was before, not merely indistinguishable from it![6]

Clearly, then, we cannot locate the self as a physical object. And here, once more, we can see the two-not-two paradox.

But what if we take the self to be some element of mind? Here again we find only **composites** of other things—all of which can be further divided. We find only a turmoil of changing feelings, conceptions, and impressions.

"But," we may ask, "doesn't **memory** hold the self together in some unified and indivisible whole? Can't we find a self in such a construction as this?" But what abiding thing, thought, feeling, or impression resides in memory? Here too we find only unceasing revision upon revision, and endless chains of additions and subtractions. The mind is no less a sea of total flux than the body.

We habitually take our abstractions for what is concrete—i.e., real. But, time after time, we find only compounded things, each of which is further comprised of compounded things. All objects are empty of their own being.

So where is this imaginary self? In the body? In the mind? The more we go hunting for a self, the more obvious it becomes that we cannot locate one.

How, then, might we go from conceiving a self to understanding that nothing actually experienced corresponds to that word? This is the crux of our problem with appearance and Reality—and our problem with dissatisfaction as well. Of the three marks of existence, "this-is-not-self" (or, more threatening, "you-are-not-self") is the truly mind-boggling one.

One reason the lack of a self is so difficult to accept is because it strikes fear in us. What could be more frightening than the thought that, more than just being a loser, "I might not even exist"? Of course, such fear is ill-founded, for if "*this* is never found to be self" **is** Reality, what would change by our learning (or accepting) this fact? It's not by Truth that we suffer, but by not *seeing what's* going on.

It's precisely because we tacitly assume (and believe in) the self that it behooves us to penetrate that fear and closely investigate immediate experience. Doing this involves going beyond paradox and confusion.

THE COMMON VIEW

I saw a poster tacked to a telephone pole. It advertised a self-help program with all the usual come-ons. It promised poise, relaxation, peace, success, and happiness. "Know yourself," it said. "Know your strengths and weaknesses," so that you may better "control your life and maximize your potential."

"Know yourself." But what the poster promised—control and all the rest—is decidedly **not** what is revealed when attempting to know the self. True study of the self reveals the self as Empty, as unable to be found. The advertised course of study, however, can only promote our everyday delusion of self—the very means by which we acquire our troubles in the first place. This "self-help" course is actually a nurturing of the desire to control our objects. It's the same old game of acquiring and manipulating—of getting and spending—but in a clever disguise, and it can lead only to misery and frustration. Instead of scrutinizing the self, it's a form of fascination with self and other. It's the road to havoc and pain.

We generally go about seeking to fulfill ourselves. Our usual approach to learning "who we are" is to investigate things "out there," which we imagine to be apart from us. Even when we turn our attention to the study of ourselves—whether it be our bodies, our minds, or our societies—we still study ourselves as though we were examining an object "out there." When we do this we become fascinated by "that out there," and soon we desire to impose our will upon "that." In short, we want to get things to go our way; we want to control things; we want to change "that" into something that will bend to our will. We do this even when our "that" is "me."

This is our usual way—but it's misery, and not at all the way to "study the self." Rather, it is to blind ourselves to Truth.

Tolstoy summed up this traditional approach quite well:

> I know that most men, including those at ease with problems of the greatest complexity, can seldom accept even the simplest and most obvious truth if it be such as would oblige them to admit the falsity of conclusions which they have delighted in explaining to colleagues, which they have proudly taught to others, and which they have woven, thread by thread, into the fabric of their lives.[7]

The study of Truth is very difficult for us, as a rule, because we are prone to discard Truth out of fear of losing the well-loved structures we've built. If anything should arise that conflicts with what we believe, even if it's a simple and obvious truth, our tendency is to ignore it and tenaciously hang on to what we've constructed our lives upon.

However, we would simply be foolish to deny Truth. If Truth forces us to rethink our position (as it surely will, simply by virtue of the fact that we are holding a position, a concept), we shouldn't begrudge that. We ought to welcome the opportunity for enlightenment.

If we defy Truth and conduct ourselves in the manner described by Tolstoy, we'll never *see* Truth. Instead we'll spin within our own ideas, our own beliefs about how things are. Like the men on Escher's stairs, unless we change how we look at things, we'll not *see* our way out.

BEING HERE

Looking "out there" for Truth, or even looking "out there" for the self, is the sort of mis-search that typifies much of our living (which is often more like striving than living). It is a form of searching under a street-lamp for the keys we lost in the woods because the light is better under the lamp.

We package the world because we believe "the light is better" when we do. We try to set things up and make life easy for ourselves, but we overlook what our actual problem is.

The real human problem has to do with how things are right *Here*, right *Now*. *This* is where we find life and death. It's not life now and death later, somewhere else. *This* is where it **all** happens. We don't die in the future; we die *Now*, in the present moment.

And so, whatever it is, we must deal with it *Here*. We cannot set it off at a distance and objectify it. We can't deal with anything if we push it off to some other place or some other time.

The Truth is that we can't go off somewhere else. There's no such place, for Reality is forever immediate and at hand.

The World simply does not function the way we imagine. All we need to verify that it doesn't is to *just see*.

Here is just *this*. Nothing ever comes and goes. We're always in the boundless room. Out of blind habit and ignorance, we move the furniture in a desperate attempt to get things just right, so that we may satisfy ourselves. Yet all the while, we never seek to inquire about the nature of the room itself.

This room, where you truly are, has no doors or windows. You'll never leave *Here*. In fact, you'll never arrive either. You're already *Here*. You can **only** be *Here*. It's because we can't leave the room, because we can't leave *Here* and *Now*, that we should study what *this immediacy* is. Only in this way can we forget ourselves and release ourselves from our immense dissatisfaction.

There's nothing mysterious here. If you want to be a pianist, or a carpenter, or a scientist, or whatever, you must pour yourself into your activity. When you do, when you really merge with your activity, you awaken into *this* room.

This is not a small, petty, me-oriented approach to life. Rather, it's to investigate *Here* and *Now*. It's *seeing* Truth. It's waking to human life.

A SELF LIKE CANTOR DUST

If we are *looking* for Truth—Reality, the way things actually are prior to our ideas about them—then we must understand that in order to *see* Truth, all that we commonly take for substantial must lose its substantiality, and all our notions of Reality must prove false.

We commonly feel "I am," but when we look for that "I," it recedes into the shadows. Still, we believe that somewhere in some dark and hidden chamber lies the homunculus of self. But if it is "there," why and where does it recede when we go looking for it? And why is there no "there" that it recedes to, where we can corner and capture it?

The World often strikes us with incredible beauty—it forever tumbles out of balance, yet all the while remains perfectly poised. We may sense this great beauty, yet what is "there" always recedes from our grasp. As we move in to investigate, our object forever recedes—right on into Emptiness.

It's like a rainbow. We may see it, yet when we try to get there it recedes from us. This is the essence of things. What things are "in themselves" forever recedes from us.

How might we think of a self that recedes as we look for it? How might we think of a self that fades, and in so doing is *seen* to identify with all it is not? Let's see if we can find an analogy.

Near the close of the nineteenth century, mathematicians began conjuring up all sorts of mathematical monstrosities that disturbed a lot of people. They were disturbing because they opened the door to constructions that "could not be," yet which clearly are. The Peano curve was one such example. Giuseppe Peano discovered a curved line that could run through every point on a plane. This disturbed mathematicians because lines are supposed to be one-dimensional. But what were they supposed to do with a line whose points could be laid down in a one-to-one correspondence to every point in a plane? Planes are two-dimensional. How could an object be both two- and one-dimensional? It was the old two-not-two paradox again.

George Cantor was no slouch at coming up with some of these mathematical monsters. For example, he created one by simply drawing a line segment, and then removing the middle third of the line. Then he removed the middle third of the remaining two segments. This process can be endlessly repeated by removing at each step the middle third of whatever line segments remain. Eventually the segments form a "dust" of points, but at each step, another step can be taken. The dust of points thus generated are infinite in number, but their total length is zero.

This may appear to be only a curious mathematical nicety at first. Such dusts, however, form perfect models for observed phenomena—including everyday phenomena. For example, in his book *Chaos*, James Gleick writes of the difficulty engineers at IBM had in getting rid of static in their telecommunications. They found that they could get rid of some of the noise by stepping up the power, but no matter how much

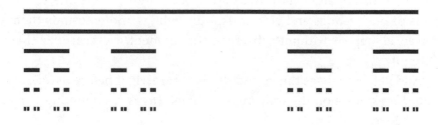

Figure 9–4. The Cantor Dust

they stepped it up, some static always remained. They needed Benoit Mandelbrot (discoverer of the Mandelbrot set) to solve their problem.

Mandelbrot was intrigued by the fact that the noise came in surges. He discovered that among periods of flawless communication were random appearances of noise. In a period of, say, an hour, there would be twenty minutes of no disturbance; then noise would reappear. When he scrutinized the bursts of noise within any one hour, he found that the bursts themselves in turn contained shorter periods of error-free communication. This pattern repeated itself on each succeedingly smaller scale. As Mandelbrot and company went looking for the periods of noise, they found at every time scale, from hours to seconds, the same repeating fractal pattern. The noise receded into Cantor dust.

Such a pattern can be recognized as being generated by eyeless Chaos, for Wholeness is involved. In such a pattern, every part is entangled with all other parts, and nothing can be extricated from the Whole.

Even our ideas and our emotions—our mental realities—elude us in this very same manner. When we attempt to grasp them, they reduce to Cantor dust. Our object is "there" if we don't scrutinize it—but as we go looking for it, it recedes. We can't get our hands on substantiality, whether the object we're looking for is a thought or a thing, or even an emotion.

Even physical substantiality, once we actually begin to examine it and try to nail it down, can't be found, as quantum physics has taught us. It **all** begins to fade away as we *look* carefully.

Every thing and every thought turns to Cantor dust the moment we seek its substantiality. The physical world, the static on the line, the objects of our feelings and emotions, our feelings and emotions themselves, all appear without substance the moment we scrutinize them carefully.

And yet, here those things and thoughts are, right before us. *Here's* a cup. *Here's* noise on the transmission lines. *Here's* the spinning Earth. *Here's* love.

This is why we must learn to examine *this*. Look at *this! This* anger! *This* love! *This* thing before me *Now*.

We usually think we have to do something about our objects, that we have to take charge and control them; but the only way to take care of Reality is simply to *see*.

It all appears substantial. Yet with love, anger, the physical world, the static on the telephone line, and so on and on, no solidity is found. Everything changes all the time. Nevertheless, *Here* we are—and *Here* is everything else.

And as we turn our light inward to illuminate the dark chamber of the self—that is, as we seek to take hold of what we've habitually taken for real—it, too, recedes even as we approach. It loses all substantiality within a lacy world of insubstantiality.

ten

⚓(TOTALITY)⚓

There was a child went forth every day,
And the first object he look'd upon, that object
he became...

—WALT WHITMAN

It is without beginning, unborn and indestructi-
ble. It is not green nor yellow, and has neither
form nor appearance. It does not belong to the
categories of things that exist or do not exist, nor
can It be thought of in terms of new or old. It is
neither long nor short, big nor small, for It tran-
scends all limits, measures, names, traces and
comparisons.

—HUANG PO

BECOMING AND FADING AWAY

Reality presents us with a tetralemma: 1) it neither is; 2) nor is not; 3) nor is both; 4) nor is neither. And so, though we haven't touched it *(what), Here* it is *(this). Here's* the noise on the line. *Here's* the book before me. *Here's* the feeling of love, the feeling of hate. But what are they? They come and go, yet nothing enters or leaves *this* room.

Our task is to *just see.* Our direct experience—i.e., perception itself— is the Undefined that says with unimpeachable authority that all things appear not in being, but in becoming and in fading away.

If we try to hold to the view that a thing is merely what it is, we'll miss it totally. We'll confuse our abstraction with *what's* Really going on. Truth reveals itself only in the moment we stop making up a story. A story is not necessary. We only make up our story out of fear.

But what can be lost by *seeing?* Truth is Truth. It will not change upon being *seen.* What changes is simply that we no longer see incorrectly.

It's only through our faulty explanations, upon which we base so much, that we suffer pain, sorrow, loss, and lamentation. And it's only through *just seeing* that we may end such suffering. The end of suffering is correct *seeing.*

However, *just seeing*—i.e., pure perception—**doesn't** mean the end of measuring, or the end of conceptualizing, or the end of discriminating consciousness. It simply means that we are no longer taken in by our objects of consciousness—by our likes and dislikes, our preferences, our goals, our desires, and our fears.

Reality is inconceivable. But still, we can *see* It. We can, and do, perceive Reality. Our problem with *just seeing* lies in learning to get past the story being told to us by others, and to ourselves by ourselves, even now. We must learn to get past conception, past paradox and confusion.

For all our wanting of things to go a certain way, there's nothing for us to do but to simply be *Here.* When "good" times come, they're *Here.* When times are "bad," they're *Here,* too. And as the "good" times slip away, if we give them too much credit, they'll grow and grow and grow in our minds, until we long to get them back. And we'll forever try to recreate those "good" times again. Thus we become spectators—not of Reality, but of what we **imagine** is Real.

We do this sort of thing endlessly, of course. What we forget is that— as bare attention bears out—the nature of good times, bad times, and everything else is to come and to go, regardless what we do, or of what we think of them.

If we don't see the "good" time for what it is, we'll not realize that it's always present, waiting to spontaneously appear. And the "bad" time too, as well as all the rest.

When we consider Reality as a Whole rather than as various collections of fragments, everything changes. Physicist David Bohm gives us an example of how the Whole appears:

> First, according to Albert Einstein's theory of relativity, the basic nature of the universe is not that of a set of interacting constituent particles. Rather, it may be described as a universal field, whose most essential quality is unbroken wholeness in flowing movement...this suggests that the whole is a primary notion, while the parts are abstractions from the whole, the traditional mechanistic notion of the constitution of the world out of separately existent parts is turned upside down.
>
> Second, the quantum theory implies that there are indivisible links of action between each object and its environment. This means that ultimately, the distinction between observer and observed, which is necessary for a mechanistic view, cannot be maintained, **not even in inanimate matter**...(even less in animate and conscious beings).
>
> Third, the whole cannot be analyzed into separate parts with preassigned interactions. Rather the whole organizes and even creates the parts. This behavior is evidently closer to organism than to mechanism.

The world created by Chaos—by Wholeness—is a terribly interesting—and satisfying, and even redemptive—world of living people, living trees, living water, living rocks, living clouds, living stars. It's a world of utter beauty. It's a world that's very easy to live in, if only we would allow it to come to us on its own terms.

When *seen* in Its Totality, like King Chaos, the World is also *seen* as dynamic peace. In other words, peace is already *Here*. It comes along with the Universe. We don't need to create it. In fact, there's nothing we can do to establish it, for it's already within the nature of the Whole.

We have a choice: we can either destroy peace, or we can let it be. Indeed, when we **try** to establish peace, we're likely to create war.

Peace is already established. Peace is already *Here* and *Now;* we need only recognize it. And once we do, we can just let it manifest.

There is no **way** to Peace. Peace is just *this,* right *Here,* before we evaluate, before we decide, before we conceive, before we pick and choose.

If we want Peace, we must simply become Peace in *this moment.*

We have no power to make the Universe live. It's alive already. We can, however, accept the Universe—the "already" Universe. Our job is to allow this Whole. This is where our power as human beings lies.

TAKING ACTION

But what about action? How are we to act in a world that is constantly changing, constantly fresh and new, and always inconceivable?

Is it even **possible** to act before we evaluate, before we decide, before we conceive, before we pick and choose? Should we even act out of our volition at all?

First of all, it's not possible to **not** act. Everything you do, say, think, and decide is an action. If you choose not to act, that choice is an action, too.

Furthermore, the whole world is changing, moving. It's in constant flux all the time. Even if you stand still in the stream, you're still interacting with all the stuff flowing around you.

Each of us has a choice: we can either act out of our confusion, out of our desires, out of our automatic responses to life's circumstances amid greed, anger, and self-delusion—or we can *just act. Just act* out of simply *seeing* the World as it arises, not as we would hope, desire, imagine, or conceive it to be.

There's a story of Mahatma Gandhi in which, as he was boarding a train, one of his sandals slipped from his foot and landed near the track. Suddenly the train began pulling away, leaving him no time to retrieve it. Immediately, Gandhi removed the other sandal and tossed it back to lie with the other along the track. When his astonished fellow passenger asked why he did this, Gandhi replied, "Now the poor man who finds it will have a pair he can use."

This story illustrates action before we decide, before we pick and choose. It is action borne of liberation—borne of utter freedom of mind. Action in compliance with the World as It becomes *this moment.*

We must learn to live by *seeing* and not by thought. We must learn to live and act from the Whole and not the part. We must learn to let the World come to us rather than thrusting ourselves upon the world.

And so, regarding that awful question, "What should I do?" coming up with an answer (as our habit demands of us) is not the point. We would only reduce Reality to concept once again. The point is simply to wake up.

Just awaken in each moment. With this you'll have right action, and so will the world.

Anything short of this, however, and you're acting out of a frozen idea—a concept. You're carrying on in defiance of Reality, as though the world were not alive, but a corpse. Such life is exquisitely painful.

It is enough to simply act out of *seeing.* When we pay attention to actual perceptions instead of our concepts—our hopes, our fears, our goals, our desires, our individual and cultural stories—then no prescription or set of commandments is necessary. Action that grows out of *seeing* is naturally responsive and appropriate to the situation, whatever it may be.

This means you're not clinging to your cherished beliefs and opinions. You're ready to toss them when you notice they don't work, when you notice they're a source of pain and anxiety.

The focus always comes back to just becoming awake—and, once again, becoming awake. This is how we free ourselves, release the world, save the environment. Just *this,* and not another idea, another best laid plan.

When we act based upon what arises in *this moment*—rather than upon what we hope, or expect, or pretend—we no longer need a script or set of guidelines to follow, because we have a clear view of Reality—

and of the Universe. We have the True evidence of direct experience to guide us.

Once we recognize *this*, we stop defying Reality.

Once you truly *see* the situation, you will act in the appropriate way. Action will take care of itself.

ENDING THE STORY

We commonly live as though at a banquet, starving—slightly aware that something's amiss. Our feeling is that we're hungry. We look about for something to eat—and, somehow, fail to *see* anything.

We don't get it.

We suffer from our thought. It tells us we must have the nugget defined. We want it defined. And it must be a jewel. A brass ring, a golden fish. It **must** be.

We want life, to be sure. But we insist on having it embalmed. Generations come and go, yet we're still taken in by the tired, old, perennial issue. We long for permanence in a world of total impermanence and relativity.

Yet if we cease to create disharmony, harmony ceases to be an issue. If we cease to create meaninglessness, the desire for meaning ceases to nag at us. If we would just stop trying to hold It in our hand, nothing would be lost.

Liberation lies in *just seeing,* in being present, in living by experience, without reliance on belief or intent, in the true freedom that is sorrow's end.

THE TRUTH

*It arrives in our lives like a table setting we don't
know how to eat with. Where is the friendly,
curved spoon, the fine-pronged fork, the fat
rounded knife?*

*We sit at the table like peasants invited to
dinner by the lord of the manor, gaping at everything we see. Here is the paté de fois gras, the
pheasant under glass, the flan in rum-caramel
sauce. We sit politely, hands in laps. Our heads
swirl with hunger. The lord smiles, gestures.
"Eat," he says, "Eat."*

—WARREN LANG

⤳(EPILOGUE)⤳

Pleasure itself teaches us, without any help from
morality or religion, that she is not what we
must seek, for as soon as we seek her, she flees.
But as soon as we drive her away, she pursues us
and seizes us with unexpected force.

— LANZA DEL VASTO

THE PARADOX OF SUFFERING

Most of us spend much of our time building barriers between ourselves and that discomfort we conceive to be "out there."

We build these barriers in many different ways, in each case attempting to transform or rearrange the world in ways we believe will satisfy us. We can have a big house in the suburbs. We can construct a fence or, better yet, a wall around our property. Maybe we can live in a neighborhood where the visible suffering of the poor is far away. If we are rich enough, we can always find a place where we don't have to view those whose suffering is readily apparent. If we have to go into places where we might encounter people living under visibly painful circumstances, we can ride around (literally or metaphorically) in a big limousine with darkened windows, so as to isolate ourselves from their suffering. We can look out if we choose to (we don't want to deny ourselves any options), but all that suffering doesn't have to peer in at us.

Most of us, however, aren't so wealthy that we can afford an actual limousine, so we set our barriers by other means. We may be a jester who makes every event into a light-hearted mock show. We may be a frail flower who regularly exacts sympathy from others. Or we may simply be a thick-skinned lout who lives like a rhinoceros, impervious to stings and stabs that would more visibly wound others.

We devise many ways to shut out these aspects of life that we don't find satisfying. But removing or isolating ourselves in some way from the world is an act of desperation: we are desperately trying to satisfy the desire to be removed from suffering.

What we don't readily understand is that (and here is the paradox) we suffer **because** we try to remove ourselves. We want somehow to isolate ourselves from pain and sorrow, from things that would wound us or reveal to us that, after all our effort, we're not really satisfied at all.

The frustration and suffering we face in trying to build barricades around ourselves results from our inability to accomplish this task. In Reality we **can't** isolate ourselves from the rest of the world. The more we attempt to build barriers between ourselves and our dissatisfaction, the more we only exacerbate our suffering.

BEYOND DISSATISFACTION

In that moment when we are together with the object of our desire, we may think or feel, "This is happiness." But then everything changes, and we lose our hold on that object. Suddenly we're unhappy. There's no way we can stop our circumstance from changing, however, for change is a mark of existence itself.

Then there are those things that we don't want to come our way. We try to escape them or avoid them, but often we can't. We didn't want to get cancer, we didn't want our son killed in an automobile accident, we didn't want a toothache, we didn't want a nuclear waste dump built in our neighborhood, we don't want to die.

But everyone's life is full of such events, in spite of what we want. Though we try to hold them off, they come to us. We always try to push away what we don't want, but no matter what we do, undesired things and circumstances come along. They're part of the fabric of life, of existence.

If we don't *see* this, we'll only become frustrated and suffer still more. We'll fool ourselves into feeling that we could have made our life more pleasurable and secure if only we were more clever, or luckier, or richer, or whatever.

We commonly try to deal with suffering by running from it, or barricading ourselves against it. **But running and putting up barriers are precisely the opposite of dealing with suffering.** Since dissatisfaction is built into the very fabric of existence itself, there's no way to escape it. We only frustrate ourselves as we try to run and to build barricades.

There's a story about a man who came to the Buddha for help. He was unhappy with his life. There was nothing overwhelmingly terrible about it, but it always presented him with an endless succession of little disappointments and complaints.

He was a farmer. And he enjoyed farming. But sometimes it didn't rain enough, or it rained too much, and his harvests were not the best.

He had a wife. And she was a good wife; he even loved her. But sometimes she nagged him too much. And sometimes he got tired of her.

And he had kids. And they were good kids. He enjoyed them a lot. But sometimes....

The Buddha listened patiently to the man's story until finally the man wound down. He looked at the Buddha expectantly, waiting for some word to fix everything. The Buddha said, "I can't help you."

The man was startled. He said, "I thought you were a great teacher. I thought you could help me."

"Everybody's got problems," said the Buddha. "In fact, we always have eighty-three problems, each one of us, and there's nothing we can do about it. If you manage to solve one problem, it's immediately replaced by another. You'll always have eighty-three problems. You're going to die, for example. For you, that's a problem, and it's one you'll not escape. We all have problems like these, and they don't go away."

The man became furious. "Then what good is your teaching?" he demanded.

"Well," said the Buddha, "it might help you with the eighty-fourth problem."

"The eighty-fourth problem?" said the man. "What's the eighty-fourth problem?"

"You don't want any problems," said the Buddha.

As to what happens to us in life, we may have little or no choice. As to how we deal with it, we have total choice.

Rather than running from our troubles, our dissatisfaction, and our suffering, rather than frustrating ourselves in a vain attempt to bring them to an end, what if we were to simply turn **toward** our problems and face them, embrace them, and in this way deal with them instead? We need not make the fact that we have problems into yet another problem.

The person who *sees* Reality is the person who embraces the suffering of others and who takes on the suffering of the world. Those who can hear the cries of others have found the secret of living. They've removed all barricades. They've stopped running away. They leave themselves wide open to what others vainly try to shut out.

But in taking on our difficulties, which must include accepting the suffering of others, we no longer have such difficulties. The world is transformed, for we no longer defy Reality. When we *see* how our wellbeing, and our suffering, are inextricably bound up with the wellbeing and suffering of others, we don't suffer—at least, not personally. We are actually embracing Truth, embracing Reality. And—here's another paradox—the full embrace of Truth doesn't register as displeasure at all. Quite the contrary; it's the only way to a life that is free of the hollow ache of meaninglessness.

There are, and have always been, ordinary people in every time and place who have learned to *see* the needs of humanity, and who take upon themselves the great suffering of the world. Some of these people who fully embraced Truth were mythologized sages (such as Buddha or Christ), but most are and were regular flesh-and-blood human beings.

If we look at these people's personal lives, we find that, even though they may have lived humbly, their lives have been rich and fulfilled.

These people do not envy others. People who feel the pain of the world feel no need to trade their lives for the common life of getting and spending. They *know* the means by which we lay waste to the Earth. And they *know* that living a life only for the immediate concerns of one-self is living the life of a failed human.

I remember hearing on the evening news some years ago about a man who had committed suicide. The note explained that he was simply "bored with the whole damn thing." He was a wealthy man, a famous man who moved through high society. He had everything that most people think they want. He "had it all"—including the ennui that comes with living a meaningless life.

But this man's story is not new. We've seen it again and again, for it's repeated endlessly. How clear it is that such a life is misery. Yet how slow are we to *just see.*

People who give themselves to others don't suffer such great misery as this. There's no ennui for them. Instead, there's the satisfaction and serenity that come from *seeing*—and living—the Truth.

We need not fear our problems. They're always with us. It's only through turning toward our dissatisfaction, and through willingly taking on suffering, that we no longer suffer. Our outward circumstance may not be pleasant, or may not appear to improve, but if we are free from the desire to be free of our difficulty, then what difficulty do we have? Thus with a clearer mind do we face our Real circumstances.

With a clearer mind we can live a life of Beauty, Awareness, and Truth, in a world that responds in kind. For those who *just see,* everything is utterly immediate and alive.

ᴡ(NOTES)ᴖ

INTRODUCTION

1. Jeremy Campbell, *The Improbable Machine,* (New York: Simon and Schuster, 1989), p. 224.

2. Ibid., p. 231.

3. James T. Cushing, "A Background Essay," in *Philosophical Consequences of Quantum Theory,* eds. James T. Cushing and Ernan McMullin, (Notre Dame, Indiana: University of Notre Dame Press, 1989), p. 17.

CHAPTER ONE

1. Einstein's term, $8\pi G/c^4 \times$ vacuum energy density, is called the "cosmological constant." Though Einstein retracted it shortly after introducing it in 1917, it has not been so easily gotten rid of [cf. *Scientific American,* 5/88, p. 106]. The fact that it is both extraneous and indispensable suggests the common occurrence of paradox when we unwittingly posit absolutes in what is utterly relative. This habit we have of unwittingly positing absolutes in our experience, and the paradoxes which result, will be explored in later chapters.

2. For further reading on this topic you may wish to turn to *Mathematics, The Loss of Certainty* by Morris Kline (Oxford University Press, 1982).

3. Roger Penrose, *The Emperor's New Mind,* (New York: Oxford University Press, 1989), p. 418.

4. Campbell, *The Improbable Machine,* p. 228.

5. W. E. Kennick, "Appearance and Reality," article in *Encyclopedia of Philosophy,* Vol. 1, edited by Paul Edwards, (New York: Macmillan, 1972), p. 135.

6. John D. Barrow, *The World Within the World,* (New York: Oxford University Press, 1988), p. 19.

7. Physicist N. David Mermin, for example, along with many others, notes that there is much to be uncertain about. Says Mermin, "certain powerful but flawed verbal and mental tools we once took for granted continue to infect our thinking in subtly hidden ways." [*Science News,* 8/5/89, Vol. 136, No. 6, p. 89] (Physicists, of course, are the people who came up with the uncertainty principle.) Physicists have good reason to be cautious these days. The bizarre realities which some of their research suggests lies behind external appearances are difficult to contemplate, to say the least. Nevertheless, by Chapter 4 we'll have managed to precisely locate just where our uncertainty—i.e., our impression of contradiction—is coming from.

8. Daniel J. Boorstin, *The Discoverers,* (New York: Vintage Books, 1985), p. 250.

9. This open and inquiring approach was adopted by the logical positivists earlier in this century. It's also used today by most scientists. As far as this approach fits into scientists' method, the method remains impeccable; but there's something profoundly wrong with science as it is commonly practiced.

Most people, scientists and logical positivists included, unwittingly weaken and ultimately completely undo their inquiry, no matter how impeccable its method, by making a few unwarranted, slyly hidden, and erroneous assumptions. I'll discuss these in some detail in later chapters.

Because both scientists and logical positivists have placed unwarranted faith in things metaphysical—i.e., things not given in direct experience—certain inevitable setbacks plague both groups. As a result, we no longer hear either group claim that Truth is what they're about—and rightly so.

10. Bernard Williams, "Descartes," article in *Encyclopedia of Philosophy,* Vol. 2, edited by Paul Edwards, (New York: Macmillan, 1972), p. 347.

11. Ibid.

12. But the word "esse"— "to be, to exist"—is still not quite reflective of actual experience, for it has a static quality about it. Since all our experience is in (or of) time, the phrase "thought is to be" does not quite hit the mark, for "to be" implies an abiding, unchanging thing. To get closer to actual experience, then, Descartes might have said "cogitatio ergo existere," or better yet, "sensus ergo existere" (consciousness, therefore becoming). The fact is, however, Truth will never be captured in words.

CHAPTER TWO

1. James T. Cushing, "A Background Essay," in *Philosophical Consequences of Quantum Theory*, eds. James T. Cushing and Ernan McMullin, (Notre Dame, Indiana: University of Notre Dame Press, 1989), p. 13.

2. Ibid.

3. Johann Rafelski, *Time*, May 15, 1989, p. 63.

4. Lys Ann Shore, "Skepticism in the Light of Scientific Literacy," in *Skeptical Inquirer*, Vol. 15, No. 1, Fall 1990, p. 4.

5. John D. Barrow, *The World Within the World*, (New York: Oxford University Press, 1988), pp. 24–5.

6. Ibid., p. 24.

7. Barrow says that the first of these beliefs saves us from solipsism, since it admits the existence of objects that do not respond to our free will. According to Martin Gardner, however, solipsism, strictly speaking, is irrefutable. Says Gardner, "There is no absolute way to refute anything outside of pure logic and mathematics." (Martin Gardner, *The Whys of a Philosophical Scrivener*, New York: William Morrow and Company, 1983, p. 13.)

But mathematics and logic can only refute what is inconsistent within a given set of assumptions. It's true that **another** set of axioms may affirm or deny our "given" set of assumptions, but then this **new** set of axioms still faces the same lack of ground as that which limited the first. Mathematics and logic are **not** conclusive of Truth but only of consistency within a given system. Experience bears this out. It's not logic, but experience, and experience alone, which must be relied upon to refute solipsism—which it does.

Solipsism **can** indeed be refuted in the same manner Descartes' *cogito* can be refuted. Experience simply does not back solipsism. In fact, contrary to Gardner's argument, solipsism is a **most** refutable proposition, if only we'd attend to bare perception. Of course, if instead we persist in maintaining our unfounded belief in an external world then we'll also persist in our tacit assumption of a self. And, as long as we hold faith in this unsubstantiated proposition—this belief—it will be just as Gardner says—we'll not refute solipsism. But there's nothing in experience that backs such a proposition. In other words, we do not need to be saved from solipsism any more than sailors of the ancient world needed saving from the perils they feared awaited them at the "edge of the Earth."

8. Gardner, p. 15 (See note 7 above).

9. Morris Kline, *Mathematics and the Search for Knowledge*, (New York: Oxford University Press, 1985), p. 21.

10. John L. Casti, *Paradigms Lost,* (New York: William Morrow and Company, 1989), p. 25.

11. Nick Herbert, *Quantum Reality,* (Garden City, New York: Anchor Press/Doubleday, 1985), p. 193.

12. Actually, though the less philosophically minded scientists—the majority—may believe they are dealing with an external world, an objective reality, and Truth, the more philosophically minded will cite William of Occam, and claim they only seek "the simple explanation." These folks might say that science is not about Truth, but about finding models that duplicate (or explain) experience. This is a more honest approach, I think. Actually, it's all science **can** do—and it's a lot.

The degree to which Occam's razor has appeared among scientists at all has come about only quite recently, with the gradual realization (by some scientists) that our models—our conceptual constructs—are, in fact, **not** Truth Itself.

13. Gardner, *The Whys of a Philosophical Scrivener,* p.15.

14. Ibid.

15. In *Philosophical Consequences of Quantum Theory* (edited by James T. Cushing and Ernan McMullin, University of Notre Dame Press, 1989), N. David Mermin introduces his Baseball Principle which illustrates the sort of problem we get into when we confuse the metaphysical (what we imagine we experience) with the empirical (what is actually experienced).

Our confusion arises due to our strong inclination to override direct experience and go with what we believe we know—i.e., with concept. Our deep habit is to not fully attend to direct perception (i.e., keep the experiment purely empirical), but rather to build upon unwarranted assumptions that result in conclusions that contradict observed facts. If we'd actually begin with what appears to perception **alone**, we'd end up with no contradiction. But we do not usually begin with perception, but with presuppositions and assumptions.

16. With the discovery of dark energy in 1998, cosmologists are now able to account for the total mass-energy of the visible universe. Dark energy accounts for 73% of it, while dark matter accounts for another 23%. That's 96% of the stuff of the universe right there—though we have no clear idea of what it is. Atomic matter, what we're familiar with in our everyday world, only accounts for about 4% of the total, with light (0.005%) and neutrinos (0.0034%) making up the rest. Putting it all together, then, the amount of positive mass-energy within a sphere of radius 13.7 billion light-years is about 10^{53} kg, which is equivalent to about 10^{11} galaxies, while the amount of negative gravitational energy

comes to about -10^{53} kg. In other words, the total mass-energy of the universe sums to zero.

17. William I. McLaughlin has made an unsuccessful attempt to set this problem to rest in his article, "Resolving Zeno's Paradoxes," in *Scientific American*, November 1994 (reprinted in 2006). As the basis of his argument, McLaughlin posits the reality of "infinitesimals," which are phantasmagorical thought constructs with no basis in perceived Reality whatsoever. McLaughlin defines such entities as "greater than zero but less than any number, however small, you could ever **conceive** [my emphasis] of writing." Ultimately McLaughlin, in words reminiscent of the Apostle Paul's references to things unseen, concludes that "an infinitesimal interval can never be captured through measurement; infinitesimals remain forever beyond the range of observation." Motion, says McLaughlin as an act of blind faith, somehow resides in these infinitesimals, and thus Zeno's paradoxes of motion are supposedly resolved.

In fact, however, since these infinitesimals are neither conceivable **nor** perceivable, they belong to neither the world of matter and energy nor even to the realms of thought or imagination. Indeed, by definition they do not (and cannot) exist at all. In short, they do not belong to Reality. This is hardly a resolution of anything, but merely a tautology masquerading as a proof.

The basic contradiction which McLaughlin's attack fails to resolve is that a thing can both change and yet remain unchanged. We'll run into this problem—this error of belief in things unseen, really—again in other guises as we consider the reality of quantum objects.

18. Herbert, *Quantum Reality*, p. 189.

19. J. S. Bell, "Six possible worlds of quantum mechanics," in *Speakable and unspeakable in quantum mechanics*, (New York: Cambridge University Press, 1987), p. 187.

20. Herbert, *Quantum Reality*, p. 40.

CHAPTER THREE

1. William Poundstone, *Labyrinths of Reason*, (New York: Anchor Press, 1988), p. 18.

2. Ibid., p. 16.

3. Strictly speaking, according to Einstein, any observer has the right to regard his or her own velocity as zero. Therefore he used the terms U and V to refer to the velocities of a projectile (U) and the vehicle (V) from which the projectile originates. The paradox stems from the fact that our commonsense view (i.e., Newton's view) dictates that the relative velocity of the projectile (relative to "me") is the sum of U+V (this is proposition A). But this contradicts

the observed fact that the speed of light is constant (proposition B). Einstein rectified this situation by replacing U+V (proposition A) with a new proposition, which states that the relative velocities of any projectile originating from any vehicle (i.e., for any value of U and V) are expressible as $U+V/1+(UV/c^2)$, where c = speed of light. This new proposition (A') does **not** contradict observation, whether our vehicle is a car or a light beam.

4. Poundstone, *Labyrinths of Reason*, p. 18.

5. Herbert, *Quantum Reality*, p. 66.

6. Ibid.

7. John Gribbin, *In Search of Schrödinger's Cat*, (New York: Bantam Books, 1984), p. 92.

8. Ibid., p. 164.

9. There are a number of books available which describe this experiment. I have relied heavily on Roger Penrose's description in *The Emperor's New Mind* (New York: Oxford University Press, 1989), pp. 231–236.

10. Penrose, *The Emperor's New Mind*, p. 235.

11. Ibid.

12. Ibid., p. 234.

13. Ibid., p. 236.

14. Ibid., p. 292.

15. Niels Bohr, quoted by Werner Heisenberg in *Physics and Beyond*, (New York: Harper and Row, 1971), p. 206.

16. Poundstone, *Labyrinths of Reason*, p. 143.

17. Nick Herbert, *Faster Than Light*, (New York: Plume Books, 1989), p. 70.

18. Ibid., p. 71.

19. J. T. Fraser, *Time: The Familiar Stranger* (Amherst: The University of Massachusetts Press, 1987).

20. Lao Tzu, *Tao Te Ching*, Chapter 11, my translation.

21. Cushing, *Philosophical Consequences of Quantum Theory*, p. 17.

CHAPTER FOUR

1. Penrose, *The Emperor's New Mind*, p. 111.

2. A tetralemma is simply a four-way, four-cornered, or "four-horned" dilemma.

3. The other two laws are: (1) the law of contradiction or, "A is not non-A," and (2) the law of excluded middle or, "everything is either A or not A." While there has been little comment regarding the law of identity (in fact, even Aristotle never stated it explicitly, though he did presuppose it), these other two laws have frequently come under attack down through the Ages. By many ac-

counts, the law of excluded middle has actually fallen. L. E. J. Brouwer and other intuitionists, for example, deny the validity of the law of excluded middle on the grounds that the truth of a proposition is not dependent upon the falsity of its contradictory. (See Morris Kline's *Mathematics; the Loss of Certainty* (Oxford, 1980), pp. 237–39.)

4. I should point out that, just because we may casually recite the phrase 1 + 1 = 2 as "one plus one is 2," it doesn't mean that the equal sign (=) symbolizes the word "is." The equal sign indicates equivalence, not an assertion of being. If you want to get closer to the phrase "one plus one is 2," you should at least use the identity sign ($\equiv$) rather than the equal sign—though this is still insufficient, since identity cannot be established among mind objects.

Please note, too, that my use of the plus symbol (+) to indicate the assertion of being (is), is not meant to imply actual being in any objective sense. We are talking strictly about mind objects here. Thus the plus symbol (+) is used only to represent the positive assertion of a singular object in the mind. Likewise, the negative symbol (−) indicates the negation of some particular mind object—which, owing to the fact that it is a mind object, involves the rest of the Universe, as we'll see.

For more details, see appendix.

5. Penrose, *The Emperor's New Mind*, p. 444.

6. Jeremy Campbell, *Improbable Machine*, p. 233.

7. Ibid., p. 234.

8. W. H. Auden wrote:

> Minus times minus is plus,
> The reason for this we need not discuss.

Auden is not the only person to be frightened by mathematical operations done with negative numbers. Such operations caused quite a stir among mathematicians in the 18th century. But it's not so difficult as all that. The greatest mathematician of that century, Leonhard Euler, justified the operation of subtracting negative numbers as being the equivalent of adding their positives because "to cancel a debt signifies the same as giving a gift." And so he also argued that to minus a minus gives us a plus, or that −1 times −1 is +1 because the product must be either +1 or −1, and since we know that 1 of −1 is −1, the negation (the −1) of −1 must be +1.

9. The standard notation for complex numbers is $a + bi$ where, if a and b are real numbers and $i = \sqrt{-1}$, the real part is a and the "imaginary" part is bi.

10. Penrose, *The Emperor's New Mind*, p. 236.

11. If you think you have seen a light wave, float around in space for a while. Unless you're near some object, you'll notice that what surrounds you appears as utter blackness. You're not seeing any light at all, even though light waves stream past you. Only when you interrupt that stream by placing an object in front of you—say, your gloved hand—is light seen (i.e., conceptualized as the image of your glove). But what appears as light (i.e., what registers in consciousness) are photons, not waves. You do not see light waves streaming from your glove to your eye. What you see are trillions of photons collectively forming a visual image (your gloved hand) in your mind. Simply put, whether were talking about live/deadness, light waves, or any other evolving superimposed quantum state—it always makes use of complex numbers, but never manifests as an object to the mind.

12. Note that it's not "wave/particle" that is comparable to "live/dead." In the case of light waves, it's "wave" all by itself that is comparable to "live/dead." In other words, as noted in the endnote above, there's no such mind object as a wave of light, just as there's no such mind object as a live/dead cat. Only when an object forms in consciousness do we have a cat (either alive or dead) or a photon (a light particle). Whether at a quantum or classical scale, prior to such formation there's no conceivable object whatsoever. Consciousness simply won't have it!

CHAPTER FIVE

1. Richard Dawkins, *The Extended Phenotype* (New York: Oxford University Press, 1999), and *The Selfish Gene* (New York: Oxford University Press, 2006).

2. Herbert, *Quantum Reality,* p. 249

3. Dr. Crypton (Paul Hoffman), "A Boy and His Brain Machine," *Science Digest,* August 1986, p. 39.

4. My discussion of Mind and conscious awareness has thus far been somewhat superficial, in that it presumes the objects of consciousness (e.g., a finger, a flower, or a star) to be real (i.e., substantial). It also presumes that these objects are "out there," apart from "me," the observer, the subject, and the ostensible possessor of conscious awareness. We thus often speak of conscious awareness as "my consciousness." As we shall see, however, in Reality no such thing is ever found. Indeed, it's only through objectless Awareness that Reality is *seen*.

"How," you may ask, "is it possible to have an objectless awareness?"

The double-slit experiment is an example. We are aware of *what's* going on, yet we cannot conceive of *what's* going on. Reality will not go into concept. Nor does it need to.

We'll look at some of the deeper aspects of conscious awareness in the next chapter. We'll also *see* that complete Awareness is subjectless as well as object-less.

5. Edmund Blair Bolles, *A Second Way of Knowing* (New York: Prentice Hall, 1991), p. xviii.

6. What we commonly call "perception" is still actually conception, in that it refers to symbolized versions of direct experience.

7. Jeremy Bernstein, *Quantum Profiles* (Princeton, NJ: Princeton University Press, 1991), p. 55.

8. I want to be clear about my use of this term, *i*. I am not saying that we cannot conceive the term itself. We know it quite well as the square root of − 1. Therefore $(i)(i) = -1$. Nothing inconceivable about that, or the many other operations we can perform with this number. What is inconceivable is any object this term is purported to designate. Whether we're talking about light streaming in waveform or Schrödinger's cat inside a sealed box, no such "objects" ever appear. Though such "objects" seem to be accounted for by the use of complex numbers, all such "objects" remain inconceivable. No one has ever seen a wave of light or a live/dead cat. But when a "photon" or a "cat" does appear in the mind (i.e., in a form accountable by real numbers alone), it is always a concept, a mind object. In fact, as we will see, **all** objects are mind objects.

9. James Gleick, *Chaos* (New York: Viking, 1987), p. 221.

CHAPTER SIX

1. Julian Jaynes, *The Origin of Consciousness in the Breakdown of the Bicameral Mind*, (Boston: Houghton Mifflin, 1990), p. 66.

2. Herbert, *Quantum Reality*, p. 249.

3. What I discuss here should not be confused with objectless Awareness, which I'll say more about later.

4. Mind, Reality, and Totality are other words for Wholeness. In Chapter 2 we considered that Wholeness, the total mass-energy of which if gathered together in one place and time, would add up to zero (see endnote 16 from Chapter 2). Consciousness, then, can be *seen* as nothing more than that function of Mind that divides what is a seamless Whole into "this" and "that." The primary division, of course, is the conscious experience of "me," and "all of that other stuff, out there, apart and distinct from me." What I refer to as *perception* in this book belongs to Mind as Whole, not to "me" as an individuated part. In other words, *perception* (unlike conception) is objectless (and subjectless). Confusingly, what we often call "perception" in everyday parlance is actually conception.

5. Penrose, *The Emperor's New Mind,* p. 406.

6. It was Gottfried Wilhelm von Leibniz who first formulated this question in 1714, in his essay "The Principles of Nature and of Grace, Based on Reason."

7. When Heisenberg discovered his principle of uncertainty, he noted that the position and the momentum of an electron, say, are represented by non-commuting matrices of numbers. In layperson's terms, this means that if the value of one matrix (i.e., set of numbers) is known, the other must necessarily remain uncertain. In other words, the path actually taken by an electron, as it moves from source to detector, cannot be pinned down.

Heisenberg assumed, rather commonsensically, that a moving electron does indeed **have** a path, but that, due to physical limitations, we simply can't measure it. His reasoning went as follows: if we try to pin down the exact position of a moving electron, we need to shine some light on it to see where it is. If we use visible light, its wavelengths are too large to detect the electron. If we use light of a higher frequency, however—say gamma rays—that light is of extremely high energy. As a result, when this light hits the electron, the impact is enough to change the electron's momentum. On the other hand, when we try to measure an electron's momentum in a similar manner, we cannot arrive at certainty about its position. This is why we cannot measure with certainty both its momentum and position—or so Heisenberg thought.

Heisenberg traveled to Copenhagen to get Bohr's appraisal of his interpretation of the meaning of quantum uncertainty. What Bohr had to say shocked Heisenberg. Bohr agreed to the principle of uncertainty, but told Heisenberg that his assumption that an electron actually takes a path is wrong. In fact, said Bohr, it's not even correct to say that an electron **possesses** position and momentum. The meaning of uncertainty in these quantum experiments, Bohr argued, is that **the very concept of a path is inherently ambiguous.**

Bohr's interpretation has been borne out by experiments. For an interesting discussion in layperson's terms of how quantum "objects" seem to take all options available to them at once, I recommend reading Richard Feynman's book, *QED* (Princeton University Press, 1985).

8. Herbert, *Quantum Reality,* p. 47.

9. Ibid.

10. Ibid., p. 48.

11. Ibid.

12. John Casti, *Paradigms Lost,* p. 489.

13. Henry P. Stapp, "Quantum Nonlocality and the Description of Nature," in *Philosophical Consequences of Quantum Theory,* p. 156.

14. Joseph Wood Krutch, *Grand Canyon*, New York: Anchor Books, 1962), p. 188.

15. Ibid., p. 189.

16. The Nature Conservancy, Minnesota Chapter, Spring 1988, p. 7.

17. As this book neared completion, I learned that Chief Seattle never actually uttered many of the sayings attributed to him. At first I decided to strike this "quote," but then I thought better of it: it's a worthy statement regardless of who made it. As the Buddha suggested, we ought to pay more attention to what is actually said than to who says it.

CHAPTER SEVEN

1. J. S. Bell's original 1964 paper, "On the Einstein-Podolsky-Rosen paradox," can be found in his *Speakable and unspeakable in quantum mechanics* (New York: Cambridge University Press, 2004), pp. 14–21.

2. Herbert, *Quantum Reality*, p. 227.

3. Francis Thompson, "Sight and Insight," in *Complete Poems of Francis Thompson* (New York: The Modern Library), p. 184.

4. Herbert, *Quantum Reality*, p. 212.

5. Ibid., p. 213.

6. Ibid., p. 229–30.

7. For a very good (and a far more detailed) description of Bell's theorem in layperson's terms, consult Nick Herbert's *Quantum Reality* (Anchor Press/Doubleday, 1985), or F. David Peat's *Einstein's Moon* (Contemporary Books, 1990). For a more technical description, I recommend *Philosophical Consequences of Quantum Theory*, edited by James T. Cushing and Ernan McMullin (University of Notre Dame Press, 1989), or "The Reality of the Quantum World," by Abner Shimony, *Scientific American*, January 1988. J. S. Bell's original 1964 paper, "On the Einstein-Podolski-Rosen paradox," as well as other related articles, can be found in Bell's *Speakable and unspeakable in quantum mechanics* (Cambridge University Press, 1987).

8. Herbert, *Quantum Reality*, p. 223.

9. Cushing, "Background Essay," in *Philosophical Consequences of Quantum Theory*, p. 10.

10. Linda Wessels, "The Way the World Isn't: What the Bell Theorems Force Us to Give Up," in *Philosophical Consequences of Quantum Theory*, p. 82.

11. Ibid.

12. Ibid.

13. Ibid., p. 94.

14. Ibid.

15. Ibid.

16. Ibid.

17. Ibid., pp. 95–6.

18. Actually, though they are optically bright sources, the light we detect is not so much visible light as it is infrared, x-ray and, in some instances, radio bands of frequency. Thus, our detection devices are not our eyes, but giant radio telescopes.

19. J. Hoover Mackin, "Concept of the Graded River," in *Geological Society of America Bulletin*, Vol. 59, No. 5, May 1948, p. 471.

20. Herbert, *Faster Than Light*, p. 191.

21. Ibid.

22. Penrose, *The Emperor's New Mind*, p. 444.

23. "The Sorites Paradox," in *British Journal for the Philosophy of Science* 20 (1969), pp. 193–202, cited by Nicholas Falletta, *The Paradoxicon* (New York: John Wiley and Sons, 1990), p. 14.

24. Don't panic. It's not necessary to know precisely what these constants mean to understand the point made here. If you wish more information on these constants, refer to Reinhard Breuer's *The Anthropic Principle* (Birkhauser, 1990).

25. This presupposes that the universe popped into existence out of nothing in a Big Bang around 13.7 billion years ago. Apart from the fact that there are insurmountable philosophical problems with this notion, it's physically impossible to look back in time and examine such an event. As we look back in time and get "close" to the Big Bang, the laws of physics themselves break down. Just for starters, there's the problem of time running more slowly as we go back in time (because the universe is getting more dense). But, of course, we can always do a thought experiment and ask, "Running more slowly for whom? For those fearless time-travelers who've gone back to those nascent times; or is that only how their world appears to us as we look back at them from the twenty-first century?"

In addition, there's the fact that, depending on your point of reference, when you look at a star 1,000 light-years away, it's either a long way away or it's right *Here*. We say it took 1,000 years for the light to get here, because that's the way it appears to us, but in the photon's world, it didn't take any time at all. So, how far away **is** "that star," in Reality? And don't forget the implications of Bell's theorem—in Reality, there is no "there," only *Here*.

There's also evidence that some particles—positrons, for example—appear to run backward in time. Do they, really? Why would anyone think that? One powerful reason is because viewing things as though this were the case paints a much simpler picture of *what* is going on overall—at least if we're okay with the fact that the inconceivable (*i*) aspect is now involved. Beyond the oddity of having some things running backward in time, it means that material in your eye interacts with material in the star all in the exact same moment, even though these "two" events appear to occur 1,000 years apart (or even 2,000 years apart, depending on how we might view the situation). I could go on.

Just how big and how old is the Universe, anyway? Do these questions even make sense, ultimately? The Big Bang is just the way It looks, that's all. There is no such event. Not *Now*.

26. Clifford M. Will, *Was Einstein Right?* (New York: Basic Books, 1993), p. 166.

27. Ibid., p. 167.

28. Shortly before his death, Einstein made this observation in a letter, dated March 15, 1955, to the family of his friend, Michele Besso, who had died the previous week.

CHAPTER EIGHT

1. Will, *Was Einstein Right?*, p. 150.

2. This is why the so-called "Higgs boson" has proven so elusive. If it has a spin of zero, it can't really form as an object of consciousness.

3. Penrose, *The Emperor's New Mind*, p. 264.

4. Jeremy Campbell, *Grammatical Man* (New York: Simon and Schuster, 1982), pp. 68–9.

5. Ibid., p. 68.

6. An interesting book, *The True and Only Heaven* by Christopher Lasch (Norton, 1991), suggests that our commonsense view of progress is askew and argues for an alternative to this "exhausted" approach to life.

CHAPTER NINE

1. An excellent description of the "quantum vacuum" phenomenon appears in "The Mystery of the Cosmological Constant," *Scientific American,* May 1988. Less technical descriptions can be found in the March 1992 issue of *Discover.*

2. For a complete, nontechnical description and discussion of the development of various cosmological models, see John Gribbin's *In Search of the Big Bang* (Bantam, 2000). Further discussions of such models can be found in *Physical Cosmology and Philosophy,* edited by John Leslie (Macmillan, 1990).

3. Our "something" can also be more generalized—for example, it can be the concept of "something" itself. That is, "something" can refer to any specified entity that somehow has (or purports or seems to have) a separate and un-

changing existence. (Bertrand Russell and others used the word "something" in this manner when they posed the question, "Why is there something rather than nothing?") In this case, the mirror opposite of "something" (the "another thing") would be "nothing"—the denial or lack of existence of some unspecified but separate, unto-itself entity. This "nothing," of course, is itself a thing, an object.

4. One way of dealing with this paradox has been to adopt a series of subscripts that allow us to make distinctions between language and metalanguage. In other words, the truth or falsity of a statement must always be evaluated by way of a higher level metastatement, designated by a higher subscript. Any statement that doesn't conform to this principle is considered neither true nor false, but ungrammatical or meaningless.

These subscripts, however, form a hierarchical chain that soon either resembles the Escher stairs (page 212), or creates an infinite regression. Thus, as we've already seen, we do not avoid the paradox by such scheming, and our situation remains ultimately just as meaningless as ever. We can only get out of this quagmire by viewing our packaging of Reality in terms of $r + i$—i.e., in terms of perception, not conception.

5. Earlier I referred to certain constants of nature. But, as I mentioned in Chapter 7, these constants are not "things"—i.e., objects of consciousness—such as ideas, coffee cups, and photons. Rather, they are indicative of ultimate opposition.

6. Penrose, The Emperor's New Mind, pp. 24–5.

7. Gleick, *Chaos*, p. 38.

A NOTE ON MATHEMATICAL ⤶(SYMBOLS AS THEY)⤷ RELATE TO MIND OBJECTS

I use mathematical symbols in this book to help clarify how mind objects are affirmed or negated in the mind. In the following examples, the number 1 represents any singular mind object.

In making the statement 1 + 1 = 2, we are simply saying: here is a mind object (1) and (+) here is another mind object (1). Taken together, these are equivalent (=) to two (mind objects). In this example, the plus sign (+), standing free of the numbers, indicates the mathematical operation of addition.

Given that this 1 represents a mind object, we also need to indicate whether the mind object is being affirmed or negated. We can indicate affirmation by rewriting the statement as: (+1) + (+1) = 2. Here the plus symbols associated with the 1s do not indicate the operation of addition, but the affirmation of the two ones—the two mind objects—on the left side of the equation.

Just as +1 indicates the affirmation of a mind object, the negative (–) of 1 is
–1. Notice, however, that the mind object (1) hasn't gone anywhere. Therefore,
–1 does not depict the absence of the mind object; rather, it indicates everything **but** the mind object, thus giving the negative of the mind object. In other
words, the negative of a mind object is everything that is not the mind object—
i.e., the rest of the Universe.

In regard to a mind object, then, what's involved when working with its negative (–1) is Totality. This is where the number i, and the mathematical function of multiplication, come into play.

We casually read $1 \times 1 = 1$ as "one times one equals one," but what we're really saying is "one of one is equivalent to one." In other words, the multiplication symbol ($\times$) represents the function "of," as in, "If you had one of one, you
would have 1."

But again, since we are dealing with mind objects, we need to indicate that
the mind object is either being affirmed or negated. We can indicate affirmation by rewriting the statement $1 \times 1 = 1$ as: $(+1) \times (+1) = +1$; or, the affirmation
of the mind object (+1) implying itself (+1) in the mind is equivalent to +1, the
affirmation of the mind object.

So:

1. A mind object implying itself in the mind can be written in
mathematical short hand as:

$$(+1) \times (+1) = +1.$$

The positive one of positive one is equivalent to positive one.

2. A mind object implying its negative in the mind can be written
as:

$$(+1) \times (-1) = -1.$$

The positive one of negative one is equivalent to negative one.

3. The negative of a mind object implying the mind object in the
mind can be written as:

$$(-1) \times (+1) = -1.$$

The negative one of positive one is equivalent to negative one.

4. The negative of a mind object implying itself in the mind can
be written as:

$$(-1) \times (-1) = +1.$$

The negative one of negative one is equivalent to positive one.

Hence, both 1 and 4, the positive implying itself and the negative implying itself—i.e., the roots of the mind object—give us the mind object.

Both 2 and 3 point to the negative of the mind object—i.e., both yield -1. But what, in implying itself, gives us -1, or the rest of Totality? In other words, what is the square root of -1?

Whatever this is, it obviously can't be a mind object. We certainly cannot conceive of it as a mind object.

Even so, though it's inconceivable, we can do what we usually do with unknowns. We can assign it a letter. In this case, we can give it the letter i, for inconceivable, since its object is inconceivable—i.e., not a mind object. i belongs to Totality.

Though we can't picture what i refers to, we can work with its mathematical properties. And when we do, we find that they neatly coincide with direct experience; with perception; with *what* is going on prior to conceptualization; and with *what* occurs before the collapse of the wave function.

With this understanding of experience, what I've referred to as "mind objects" can now be *seen* as manifestations of Mind. And Mind Itself is *seen* as none other than Totality. In other words, there is nothing apart from Mind.

SELECTED BIBLIOGRAPHY

Abbot, Larry. "The Mystery of the Cosmological Constant." *Scientific American*, May 1988, pp. 106–13.

Barrow, John D. *The World Within the World*. New York: Oxford University Press, 1988.

Bell, J. S. *Speakable and unspeakable in quantum mechanics*. New York: Cambridge University Press, 2004.

Bernstein, Jeremy. *Quantum Profiles*. Princeton: Princeton University Press, 1991.

Blofeld, John. *The Zen Teaching of Huang Po: On the Transmission of Mind*. New York: Grove Press, Inc., 1984.

Bohm, David. *Wholeness and the Implicate Order*. London: Ark Paperbacks, 1984.

Bolles, Edmund Blair. *A Second Way of Knowing: The Riddle of Human Perception*. New York: Prentice Hall Press, 1991.

Boorstin, Daniel J. *The Discoverers: A History of Man's Search to Know His World and Himself*. New York: Vintage Books, 1985.

Breuer, Reinhard. *The Anthropic Principle: Man as the Focal Point of Nature*. Boston: Birkhäuser, 1990.

Briggs, John and Peat, F. David. *Turbulent Mirror: An Illustrated Guide to Chaos Theory and the Science of Wholeness*. New York: Harper and Row, 1989.

Campbell, Jeremy. *Grammatical Man: Information, Entropy, Language and Life*. New York: Simon and Schuster, 1982.

Campbell, Jeremy. *The Improbable Machine: What the Upheavals in Artificial Intelligence Research Reveal About How the Mind Really Works*. New York: Simon and Schuster, 1989.

Casti, John L. *Paradigms Lost*. New York: William Morrow and Company, 1989.

Chang, Garma C.C. *The Buddhist Teaching of Totality: The Philosophy of Hwa Yen Buddhism.* University Park: The Pennsylvania State University Press, 1977.

Churchland, Paul M. and Churchland, Patricia Smith. "Could a Machine Think?" *Scientific American,* January 1990, pp. 32–37.

Cleary, Thomas. *Entry Into the Inconceivable: An Introduction to Hua-yen Buddhism.* Honolulu: University of Hawaii Press, 1995.

Conze, Edward. *Buddhist Thought in India: Three Phases of Buddhist Philosophy.* Ann Arbor: University of Michigan Press, 1973.

Cook, Francis H. *Hua-yen Buddhism: The Jewel Net of Indra.* University Park: The Pennsylvania State University Press, 1977.

Crutchfield, James P.; Farmer, J. Doyne; Packard, Norman H.; and Shaw, Robert S. "Chaos." *Scientific American,* December 1986, pp. 46–57.

Cushing, James T. and McMullin, Ernan, eds. *Philosophical Consequences of Quantum Theory: Reflections on Bell's Theorem.* Notre Dame, Indiana: University of Notre Dame Press, 1989.

Dawkins, Richard. *The Extended Phenotype.* New York: University of Oxford Press, 2006.

Dogen Zenji. *Shobogenzo: Zen Essays by Dogen.* Translated by Thomas Cleary. Honolulu: University of Hawaii Press, 1986.

Eckel, Malcolm David, trans. *Jñanagarbha's Commentary on the Distinction Between the Two Truths.* Albany: State University of New York Press, 1987.

Falletta, Nicholas. *The Paradoxicon.* New York: John Wiley and Sons, 1990.

Feynman, Richard P. *QED: The Strange Theory of Light and Matter.* Princeton: Princeton University Press, 1985.

Gardner, Martin. *The Whys of a Philosophical Scrivener.* New York: William Morrow and Company, 1983.

Gleick, James. *Chaos: Making a New Science.* New York: Viking, 1987.

Grave, A. S. "Common Sense." Article in *Encyclopedia of Philosophy,* Vol. 2, ed. by Paul Edwards. New York: Macmillan, 1972.

Gribbin, John. *In Search of Schrödinger's Cat.* New York: Bantam Books, 2000).

Hawking, Stephen W. *A Brief History of Time.* New York: Bantam Books, 1998.

Herbert, Nick. *Faster Than Light: Superluminal Loopholes in Physics.* New York: Plume Books, 1989.

Herbert, Nick. *Quantum Reality: Beyond the New Physics.* Garden City, New York: Anchor Press/Doubleday, 1985.

Hirst, R. J. "Realism." Article in *Encyclopedia of Philosophy,* Vol. 7, ed. by Paul Edwards. New York: Macmillan, 1972.

Inada, Kenneth K. *Nagarjuna: A Translation of his Mulamadhyamakakarika with an Introductory Essay.* Tokyo: The Hokuseido Press, 1970.

Jaynes, Julian. *The Origin of Consciousness in the Breakdown of the Bicameral Mind.* Boston: Houghton Mifflin, 1990.

Kalupahana, David J. *Nagarjuna: The Philosophy of the Middle Way.* Albany: State University of New York Press, 1986.

Kennick, W. E. "Appearance and Reality." Article in *Encyclopedia of Philosophy,* Vol. 1, ed. by Paul Edwards. New York: Macmillan, 1972.

Kline, Morris. *Mathematics, the Loss of Certainty.* New York: Oxford University Press, 1980.

Kline, Morris. *Mathematics and the Search for Knowledge.* New York: Oxford University Press, 1985.

Krutch, Joseph Wood. *Grand Canyon.* Garden City, New York: Anchor Books, 1962.

Leslie, John, ed. *Physical Cosmology and Philosphy.* New York: Macmillan, 1990.

Masao, Abe. "Dogen on Buddha Nature." *Eastern Buddhist.* Vol. IV, No. 1, May 1971, pp. 28–71.

Minsky, Marvin. *The Society of Mind.* New York: Simon and Schuster, 1986.

Moore, G. E. *Some Main Problems of Philosophy.* New York: Collier Books, 1966.

Moser, Paul K. and vander Nat, Arnold, eds. *Human Knowledge: Classical and Contemporary Approaches.* New York: Oxford University Press, 1987.

Padmasambhava. *Self-Liberation Through Seeing With Naked Awareness.* Translated by John Myrdhin Reynolds. Barrytown, New York: Station Hill Press, 1989.

Peat, F. David. *Einstein's Moon: Bell's Theorem and the Curious Quest for Quantum Reality.* Chicago: Contemporary Books, 1990.

Peitgen, H.-O. and Richter, P. H. *The Beauty of Fractals: Images of Complex Dynamical Systems.* Berlin: Springer-Verlag, 1986.

Penrose, Roger. *The Emperor's New Mind.* New York: Oxford University Press, 1989.

Peterson, Ivars. "Quantum Baseball: A Baseball Analogy Illuminates Paradox of Quantum Mechanics." *Science News,* August 5, 1989, Vol. 136, No. 6, pp. 88–9.

Prebish, Charles S., ed. *Buddhism: A Modern Perspective.* University Park: The Pennsylvania State University Press, 1975.

Poundstone, William. *Labyrinths of Reason: Paradox, Puzzles and the Frailty of Knowledge.* New York: Anchor Press, 1988.

Quinton, Anthony. "Knowledge and Belief." Article in *Encyclopedia of Philosophy,* Vol. 4, ed. by Paul Edwards. New York: Macmillan, 1972.

Reynolds, John Myrdhin, trans. *Self-Liberation Through Seeing With Naked Awareness.* Barrytown, N.Y.: Station Hill Press, 1989.

Rock, Irvin. *Perception.* New York: Scientific American Library, 1984.

Rothman, Tony. "The Seven Arrows of Time." *Discover,* February 1987, pp. 63–77.

Rothman, Tony. "A 'What You See Is What You Beget' Theory." *Discover,* May 1987, pp. 90–9.

Searle, John R. "Is the Brain's Mind a Computer Program?" *Scientific American,* January 1990, pp. 26–31.

Sextus Empiricus. *Outlines of Pyrrhonism.* Translated by R. G. Bury. Buffalo: Prometheus Books, 1990.

Shimony, Abner. "The Reality of the Quantum World." *Scientific American,* January 1988, pp. 46–53.

Sprung, Mervyn. *Lucid Exposition of the Middle Way: The Essential Chapters from the Prasannapada of Candrakirti.* Boulder: Prajña Press, 1979.

Stcherbatsky, Th. *The Central Conception of Buddhism.* Delhi: Motilal Banarsidass, 1974.

Stenger, Victor J. "The Spooks of Quantum Mechanics." *Skeptical Inquirer,* Vol. 15, No. 1, Fall 1990, pp. 51–61.

Stewart, Ian. *Does God Play Dice?: The Mathematics of Chaos.* New York: Basil Blackwell Ltd., 1989.

Streng, Frederick J. *Emptiness: a Study in Religious Meaning.* Nashville: Abingdon Press, 1967.

Stroll, Avrum. "Identity." Article in *Encyclopedia of Philosophy*, Vol. 4, ed. by Paul Edwards. New York: Macmillan, 1972.

Stough, Charlotte. *Greek Skepticism: A Study in Epistemology*. Berkeley: University of California Press, 1969.

Thurman, Robert A. F. *The Central Philosophy of Tibet: A Study and Translation of Jey Tsong Khapa's Essence of True Eloquence*. Princeton: Princeton University Press, 1991.

Tsang, Hsüan. *Doctrine of Mere-Consciousness (Ch'eng Wei-Shih Lun)*. Translated by Wei Tat. Hong Kong: Dai Nippon Printing Co., Ltd., 1973.

Will, Clifford M. *Was Einstein Right?: Putting General Relativity to the Test*. New York: Basic Books, 1993.

Wright, Robert. "Did the Universe Just Happen?" *The Atlantic*, April 1988, pp. 29–44.

⼳(INDEX)ᴍ

ABOUT THE AUTHOR

Steve Hagen has been a student of Buddhist thought and practice since 1967. He became a student of Dainin Katagiri Roshi in 1975, and continued on to be ordained a Zen priest by Katagiri Roshi in 1979. He has studied with teachers in the U.S., Asia, and Europe, and in 1989 received Dharma transmission (the endorsement to teach) from Katagiri Roshi.

Steve founded the Dharma Field Meditation and Learning Center (www.dharmafield.org) in Minneapolis, Minnesota, where he lives. He currently maintains an active role at the center, where he leads classes, meditations, sesshins and more. He has written four books that help to clarify Buddhism, and his *Buddhism Plain and Simple* is among the bestselling books on the subject.

Sentient Publications, LLC publishes books on cultural creativity, experimental education, transformative spirituality, holistic health, new science, ecology, and other topics, approached from an integral viewpoint. Our authors are intensely interested in exploring the nature of life from fresh perspectives, addressing life's great questions, and fostering the full expression of the human potential. Sentient Publications' books arise from the spirit of inquiry and the richness of the inherent dialogue between writer and reader.

Our Culture Tools series is designed to give social catalyzers and cultural entrepreneurs the essential information, technology, and inspiration to forge a sustainable, creative, and compassionate world.

We are very interested in hearing from our readers. To direct suggestions or comments to us, or to be added to our mailing list, please contact:

SENTIENT PUBLICATIONS, LLC

1113 Spruce Street
Boulder, CO 80302
303–443–2188
contact@sentientpublications.com
www.sentientpublications.com